CHROMOSOME MANIPULATIONS
AND
PLANT GENETICS

A supplement to *Heredity*, Vol. 20 1965

CHROMOSOME MANIPULATIONS
and
PLANT GENETICS

The contributions to a symposium
held during the Tenth International Botanical Congress
Edinburgh 1964

Edited by
RALPH RILEY
Plant Breeding Institute, Cambridge

and

K. R. LEWIS
Botany School, Oxford

Springer Science+Business Media, LLC

1966

First published 1966

ISBN 978-1-4899-6259-1   ISBN 978-1-4899-6561-5 (eBook)
DOI 10.1007/978-1-4899-6561-5

Originally published by Plenum Press in 1966.
Softcover reprint of the hardcover 1st edition 1966

PREFACE

THE symposium, at which the papers collected together in this volume were delivered, formed part of the programme of the Section for Cytology and Genetics of the Tenth International Botanical Congress which assembled in Edinburgh in August 1964. It was intended in the Symposium to survey some of the recent contributions to the understanding of genetical, cytological and evolutionary problems made by conscious adjustments of the chromosome complements of plants. However, while this was the prime concern, since all the species considered were either crop plants or their relatives, the matters discussed were clearly relevant to the genetic improvement of the agricultural potentialities of the crops involved.

From the first establishment of the chromosome theory of heredity, information derived from the study of nuclear cytology has contributed to the understanding of genetics. Indeed the intimate relationships of the two disciplines led to such adjustments in the conceptual framework of both that the area of contact has for long been called *cytogenetics*. Cytogenetics consists, therefore, of the use of chromosomal techniques to obtain genetical information and the use of genetics to illuminate chromosomal behaviour; and consequently of the study of the interrelationships of cytology and genetics.

Many of the classic techniques of cytogenetics have employed organisms with specifically created, or selected, deviant chromosomal conditions. For example, the matching of the genetical and cytological linkage maps in *Drosophila* depended upon both the genetic effects and the salivary gland pairing of chromosomes in deficiency, translocation and inversion heterozygotes; while in corn—the classic species in higher-plant genetics—the co-ordination of the cytological and genetical maps was made possible by the use of aneuploids, by the observation of pachytene pairing in translocation heterozygotes, and by the determination of recombination values between genetic markers and translocation exchange points.

In these and many other organisms, both plant and animal, cytogenetics has occupied a central position in the development of our understanding of inheritance. During the past decade, aneuploid studies, by demonstrating new principles in the determination of sex and by other contributions to our knowledge of the genetics of man, have shown that cytogenetics continues to play an important role in animal genetics.

New possibilities for cytogenetics can also be seen in the investigation of higher plants. In the main these are not concerned with the detailed organisation of chromosomally localised activities, although an exception to this may be seen in some aspects of the investigation of the regulation of meiotic chromosome pairing. Perhaps, however—to continue the digression—it is not surprising that the study of meiotic behaviour, especially chromosome pairing—with all that this process implies for the understanding of genetics—is most readily conducted on plants and by cytogenetic procedures. Certainly over recent years the use of cytogenetic material has given results leading to the formulation of new concepts of the genetic determination of meiotic behaviour.

Generally, however, the major changes in chromosome complements, with which cytogenetics is usually concerned, are better suited to the investigation of the integration of the total genotype. The work on the potato and the tomato, described in the papers that follow, illustrates this rather well; but it is perhaps best exemplified by work on the details of the genetic architecture of wheat—particularly that of Sears. From this and similar work the internal adjustment and balanced organisation of the genotypes of polyploids become more readily discernible. The importance of this should not be underestimated since an appreciation of the co-ordinated functioning of the total genotype is as important to those interested in inheritance in plants—whether as synthesising breeders or as analysing geneticists—as is detailed information about the structure and operation of particular loci. One means of producing such an appreciation of the overall organisation of the genotype is provided by cytogenetics.

Indeed it is fair to say that our improved comprehension of the genetics of higher plants, particularly of polyploid species, is owed in considerable measure to the contributions of cytogenetics. It was for this reason that the symposium was organised, and while the emphasis on economic plants is understandable—indeed some would hold desirable—it is of interest to note that the narrow botanical range encompassed contrasts with a commendable breadth of treatment. From this an optimistic assessment can be formulated of the future contributions of chromosome studies to many genetical problems—both pure and applied—encountered in plant material.

RALPH RILEY.

Plant Breeding Institute,
Cambridge.

CONTENTS

SOME EXPERIMENTAL APPLICATIONS
OF ANEUPLOIDY IN *NICOTIANA*

D. R. CAMERON

Department of Genetics, University of California, Berkeley 4, California, U.S.A.

1. INTRODUCTION

ALTHOUGH chromosomal deviants of all types occur sporadically in *Nicotiana*, our experimental procedures, for reasons of expediency, have largely dealt with monosomics and polyploids. In the former, studies have usually been concerned with problems of a fundamental nature, but there have been practical implications as well. The monosomic collection was developed over a period of about twenty years, almost entirely through the painstaking efforts of the late Roy E. Clausen. Some monosomics arose spontaneously, in the progeny of other monosomics as a rule, some were obtained by hybridising *N. tabacum* (*tbc*) and *N. sylvestris* (*slv*) followed by backcrossing, and one was derived by the use of the genically controlled asynaptic condition. This last method would have been more useful, no doubt, had it not been for the dearth of suitable markers in our stocks. For many years we have propagated 24 monosomic lines, corresponding to the haploid complement of *tbc*, but as we shall see, there are still some complications that require explanation. I propose to discuss some of the features associated with monosomic analysis, as well as developments resulting from other aspects of aneuploidy.

2. MONOSOMIC ANALYSIS

(i) *Genetic mutants*

Of the varied genetic uses for which a set of monosomics is valuable, the primary one is the association of mutant loci with specific chromosomes. In *Nicotiana* this procedure has demonstrated the chromosomal relations of about 40 dominants and recessives, a few of the tests having been inconclusive. Some chlorophyll-deficient types, which give excellent Mendelian ratios in a background of modern commercial varieties, fail to give positive results in monosomic tests as a consequence of modifying genes present in monosomic Red Russian tobacco (Povilaitis and Cameron, 1963). It is to be hoped that this may be overcome by a few generations of backcrossing. In this programme our aim has been to obtain at least one genetic marker for each of the 24 chromosomes, but so far the distribution has been far from random. Seven of the 24 have failed to reveal any association of gene with chromosome, while six carry three or more loci each.

(ii) Translocations

The monosomic series has also proved of value in identifying the chromosomes involved in translocation complexes in hybrids that bring together contributions from primitive races with those of commercial varieties. Fourteen different reciprocal translocations, in homozygous condition, have been recognised, and each has been hybridised with the monosomics to determine the specific chromosomes which entered into the exchanges. Most of the races which differed from " standard " were identical in structure, whereas the other 13 types were characteristic of only 15 races. That is, in nearly all instances a particular translocation type was associated with a single local race. To date, 13 of the 24 chromosomes have been shown to participate in the various interchanges. One would expect naturally occurring translocations to arise by exchanges between homoeologous members of the complement, as is indeed the situation in wheat (Riley and Kempanna, 1963). On the contrary, in our material, all the 4-complexes comprise two chromosomes of either *slv* or *tmn* (*N. tomentosa*) derivation rather than one of each. In one such hybrid, only one of the two chromosomes could be identified—a member of the *slv* genome. The second was partially homologous to the S-chromosome which, in monosomic condition, regularly reveals a high frequency of trivalents at MI in PMC. So it would be particularly desirable to identify this member of the ring-of-4. A similar problem has arisen during attempts to localise both loci involved in duplicate factor situations where one chromosome was readily identified leaving the other in doubt.

(iii) Genetic constitution of varieties

It was earlier demonstrated (Clausen and Cameron, 1950) that the white seedling character in *tbc* was controlled by duplicate genes, although in Red Russian tobacco F_2 ratios conformed to monogenic expectation. The assumption was made that a similar situation may exist for a number of other character-contrasts as a consequence of the amphidiploid ancestry of this species. An example which seemed to lend itself to this type of investigation was normal *versus* mammoth growth differential. This has been repeatedly described as an instance of simple Mendelian heredity, the *mm* gene having been assigned to the F-chromosome. Yet in hybrids between *tbc mm* and both *slv* and *tmn*, the characteristic photoperiodic response had been suppressed. Moreover, F_2 cultures from an intervarietal hybrid involving Red Russian and a Peruvian accession included mammoth plants.

To test the hypothesis that duplicate factors were indeed responsible, a set of 25 varieties was crossed with haplo-F Red Russian. Selected haplo-F plants in the progenies were then test-crossed to mammoth, and the resulting populations scored as to growth habit and whether monosomic or disomic. Mammoth plants usually do not flower early enough in the fall for classification of the last two categories, but the genetic constitutions with regard to the *mm* loci could be determined on the basis of percentages of *mm* individuals present. Excluding one test that gave anomalous results, the following conclusions were drawn:

Number of varieties	Percentage normal	Percentage mammoth	Constitution
16	27·3	72·7	*Mm' mm"*
5	73·3	26·7	*Mm' Mm"*
3	50·3	49·7	*mm' Mm"*

Varietal studies dealing with other recessive markers were less convincing, on account of difficulties in scoring when unrelated genetic backgrounds were combined in the hybrids.

(iv) *Interspecific gene transfer*

More recently, considerable attention has been directed to the nature of the incorporation of chromosomal material into *tbc* from various other species. An extensive backcross programme utilising recessive markers of *tabacum* has culminated in the establishment of lines with 24 bivalents having corresponding dominant loci from such species as *slv*, *glutinosa* (*glt*) and *plumbaginifolia* (*pbg*). The objective was to determine whether introgression had been preferential or whether it had been a random process. In addition, the tests might well give an indication of species relationships where species not ancestral to *tabacum* were under study. A preliminary account of this work has appeared (Cameron, 1962), but since that time a clearer picture of the situation has begun to emerge.

Limitations of time and space have precluded conventional monosomic analysis of the introgressed lines of *tbc*. Instead, plants monosomic for the chromosomes known to bear the corresponding recessives were pollinated, using the introgressed derivatives, for example haplo-F (*Mm*) × *tbc slv Mm*. Selected monosomic plants among the progenies were then crossed, using pollen from the appropriate recessive *tbc*. The presence of four classes in each of the resulting cultures showed that the introduced chromosomal material had been incorporated in a chromosome other than the bearer of the locus concerned. Only two classes were observed (monosomic recessive and disomic dominant) when the corresponding chromosome was the recipient. The tests did not reveal whether the recessive locus actually had been replaced by the alien segment.

There were a few instances where individual exceptional plants occurred, and there is, as yet, no satisfactory explanation for them. In general, the results were clear-cut and may be summarised as follows:

(i) The C-chromosome (*wh*) is a member of the *tmn* genome and was the recipient when the dominant (*Wh*) was introduced from *tmn*.

This was not the result when *N. otophora* (*oto*) or *N. setchellii* (*stc*) was the Tomentosae species contributing the non-white region. This finding was unexpected, in view of the high degree of homology expressed in hybrids among members of this section of the genus. In a second test involving a *tmn* monosome (*G*) and the non-tinged locus transferred from *tmn*, it was again found that the *tmn* chromosome had been the recipient.

(ii) *N. glutinosa* has been included in the Tomentosae section also, but on the basis of morphology and chromosome behaviour it is much less closely related than are other members of the group. It was, therefore, surprising to find that, in both tests wherein this species served as donor, the results were as would be expected if the *glt* genome were fully homologous with the *tmn* genome of *tbc*. This was fore-shadowed by the work of Gerstel (1948), in which transfer of the necrotic type of resistance to mosaic was associated with the H-chromosome of *tbc*.

TABLE 1

Chromosomal associations of introgressed loci

Monosome	Donor species and locus	Monosomic		Disomic	
		dominant	recessive	dominant	recessive
C	*stc Wh*	11	20	31	16
C	*tmn Wh-P*	0	29	13	0
F	*glt Mm*	0	41	44	0
F	*slv Mm*	12	*	11	*
G	*tmn Tg*	0	4	44	0
P	*glt Fs*	5	2	43	40
P	*slv Fs*	0	2	36	0

*44 *mm* plants which flowered too late for classification. *stc* = setchellii; *tmn* = tomentosa; *glt* = glutinosa; *slv* = sylvestris; *Wh* = non-white; *Wh-P* = non-white, pale; *Mm* = non-mammoth; *Tg* = non-tinged; *Fs* = non-fasciated.

(iii) The P-monosome is a member of the *slv* genome of *tbc* and carries a recessive gene controlling fasciation. Two tests were studied where *slv* had provided dominant loci and again results accorded with expectation. In the non-fasciated series, the P-chromosome had been the recipient, whereas in a test involving *slv* and the non-mammoth locus, the F-chromosome, being in the *tmn* genome, had not participated in the exchange. Some of these relationships are set forth in table 1. For a complete evaluation of the situation much more extensive data will be required.

3. CHROMOSOMAL ADDITION ANEUPLOIDS

The effects of alien chromosomes when added to the complement of *tbc* were intriguing and some of them have been published (Ar-rushdi, 1957; Cameron and Moav, 1957; Moav, 1961). One of the problems awaiting solution is the mechanism of stabilisation in variegated lines derived from addition of chromosomal material from such species as *tmn* or *pbg*. It is this feature I should like to consider now.

Several years ago, two pedigrees were initiated, one involving *wh* flower colour and the other *tg*. Tetraploid *wh* and *tg tbc* plants were pollinated by *pbg* and the resulting sesquidiploids backcrossed to diploid *tbc* having the recessive traits. In both lines the plants so derived were *wh* or *tg* and variegated (*Vg*) with carmine and *wh* or *tg* streaks on the corollas. In addition, some of the *tg* populations included a few plants with self-coloured (*Sc*) flowers. In the succeeding generation, some individuals had already reverted to the 24-paired condition and in these cultures the percentage of self-coloured plants rose to about 50. In progenies derived from selfing *Sc* plants, no further variegation was observed. Parallel populations maintained by crossing *Vg* × *tg* continued to yield a small number of *Vg* plants in each generation and these were characterised by a constitution of 23″+2′. They also occasionally produced plants with self-coloured flowers. In the second generation following return to 24″, two populations consisted of self-coloured plants only, and apparently these lines are fully stabilised.

In the *wh* lineage, the results were somewhat different. As before, a 24″ condition was attained quite rapidly, but when such plants having coloured flowers were selfed, the ensuing progenies continued to segregate *Sc* and *wh* in frequencies varying from 50 to 75 per cent. Only after six generations of selfing did an exclusively *Sc* population appear, and, as in the *tg* series, the line continued to breed true. Selected plants of this type were crossed with *Sc* plants from segregating cultures, and in each instance the derived progenies were all *Sc*. The same result was obtained using similar crosses in the *tg* series, and when F_2 cultures were studied in both, some were found to consist solely of *Sc* plants while others again segregated. Unfortunately, the number of populations which could be grown was very small.

4. DISCUSSION

Some of the beneficial aspects and difficulties associated with the use of aneuploids in solving genetic and evolutionary problems have been considered. The implications, in most of the situations described, are quite clear. Concerning the variegation series, however, further comment may be in order. One puzzling feature is the apparent difference between the *wh* and *tg* lineages in the rapidity with which stabilisation was achieved. This might be attributed to differential transmission frequencies for different *pbg* chromosomes, but previous studies have indicated a rather constant value for transmission of alien chromosomes, an exception being pollen killer. Another problem concerns the failure of *Vg* to occur in progenies of *Sc* plants, although both types are produced when *Vg* individuals are selfed or backcrossed. It is also difficult to explain the constancy of the 23″+2′

constitution of Vg plants once this level is reached. In early generations Vg plants could be $24''+1'$, but none of this constitution has been encountered among descendants of plants with $23''+2'$.

Apparently, variegation results when a pbg chromosome is present either as an extra chromosome or, when unpaired, in association with a tbc univalent. The infrequent Sc plants in progenies from Vg could then be homozygous substitution products. A reasonable explanation for the origin of Sc plants may be the occurrence of an exchange of chromosomal material between the two species. Subsequently, segregation continues as a result of the presence of an unmodified tbc chromosome which pairs with the interchange product. If pollen carrying the altered chromosome is at a disadvantage in competition with grains possessing the normal tbc contribution, most or all of the resulting Sc plants would be heterozygous, and these were the ones which were selected for propagating the Sc lines. Whenever selection involved a homozygous plant that had resulted from one of the infrequent fertilisations by pollen having the altered chromosome, an exclusively Sc culture was the result. Once homozygosity was attained, the line bred true. Some of these possibilities are amenable to experimental test, and additional information should be forthcoming in due time.

5. SUMMARY

1. Some of the applications and difficulties associated with the use of aneuploids in cytogenetic studies have been reviewed. It was pointed out that available genetic markers are distributed very unevenly on the chromosomes of $N.$ $tabacum$.

2. In contrast with the situation in wheat, both chromosomes involved in various simple translocation complexes have been derived from one or other of the ancestral species of tbc.

3. It has been established that the mammoth growth character contrast depends upon duplicate factors. Varieties were analysed in which all three possible genotypes occurred.

4. Studies of introgressed loci in tbc revealed that chromosome exchange was based mainly on degrees of homology. On the other hand, an unrelated species such as $glutinosa$ may participate more readily in gene transfer than some of the more closely related species.

5. Variegation from alien chromosome addition has been considered some in detail, especially in connection with the derivation of stable self-coloured populations.

6. REFERENCES

AR-RUSHDI, A. H. 1957. The cytogenetics of variegation in a species hybrid in $Nicotiana$. $Genetics$, 42, 312-325.

CAMERON, D. R. 1962. Studies of introgressed loci in $N.$ $tabacum$. Tobacco, $Tobacco$ $Science$, 6, 137-139.

CAMERON, D. R., AND MOAV, R. 1957. Inheritance in $Nicotiana$ $tabacum$. XXVII. Pollen killer, an alien genetic locus inducing abortion of microspores not carrying it. $Genetics$, 42, 326-335.

CLAUSEN, R. E., AND CAMERON, D. R. 1950. Inheritance in *Nicotiana tabacum*. XXIII. Duplicate factors for cholorophyll production. *Genetics, 35,* 4-10.

GERSTEL, D. U. 1948. Transfer of the mosaic-resistance factor between H chromosomes of *Nicotiana glutinosa* and *N. tabacum*. *Jour. Agric. Res. 76,* 219-223.

MOAV, R. 1961. Genetic instability in Nicotiana hybrids. II. Studies of the Ws (pbg) locus of *N. plumbaginifolia* in *N. tabacum* nuclei. *Genetics, 46,* 1069-1087.

POVILAITIS, B., AND CAMERON, D. R. 1963. A mutation causing chlorophyll deficiency in *Nicotiana tabacum*. *Canad. J. Genet. Cytol., 5,* 233-238.

RILEY, R., AND KEMPANNA, C. 1963. The homoeologous nature of the non-homologous meiotic pairing in *Triticum aestivum* deficient for chromosome V (5B). *Heredity, 18,* 287-306.

CHROMOSOME ENGINEERING IN *LYCOPERSICON*

CHARLES M. RICK and GURDEV S. KHUSH
University of California, Davis, California, U.S.A.

1. INTRODUCTION

THE purpose of this article is to summarise the present state of knowledge concerning chromosomal manipulations that are possible in the genus *Lycopersicon*. References to the pertinent literature will be given, and examples presented.

The garden tomato (*L. esculentum*) offers many advantages for cytogenetic investigations. Its flower structure ensures automatic self-pollination, yet precisely controlled hybridisations are feasible on a large scale. A vast array of monogenic mutants—estimated now in the neighbourhood of 500—is available, and more than 150 have been placed in their respective linkage groups. The 12 pairs of chromosomes can be distinguished from each other in their extended condition at pachytene by their relative lengths, centromere positions and distributional patterns of heterochromatin. The garden tomato can be hybridised with all known tomato species, and even with *Solanum pennellii*, to yield vigorous hybrids of sufficient fertility to permit unlimited genic transfers. As a common denominator for the genus, its favourable genetic traits can be utilised to investigate many aspects of interspecific relations. The large-scale culture of the tomato as an economic plant permits a fortunate exchange of materials and ideas between breeders and geneticists.

The limits of chromosomal unbalance tolerated by the tomato, though narrow, allow many cytogenetic manipulations that elucidate the nature of the genome. They will be considered in order according to the nature of the chromosome modifications.

2. TRISOMICS

(i) *Primary trisomics*

The number of whole chromosomes that can be added to the tomato genome within the limits of viability is very small. Lesley (1928, 1932) and Rick and Barton (1954) showed that the cultivar San Marzano tolerates a maximum of three extra chromosomes and that even a single extra chromosome reduces vigour and fertility and causes profound phenotypic modifications, characteristic for each primary trisomic type. Subsequently a markedly higher tolerance and lesser phenotypic effect was found in derivatives of a sesquidiploid hybrid of *L. esculentum* × *L. peruvianum* (Soost, 1958) and in the primitive cultivar Red Cherry (Rick and Notani, 1961).

The extra chromosome of each primary trisomic has been identified by its morphological landmarks in pachytene. Fertility and transmission of the extra chromosome are sufficient to permit inheritance studies with each primary, although special techniques are needed to cajole certain ones to set and mature fruits. The inheritance of marker genes in the progeny of trisomic heterozygotes was studied for evidence of modified ratios that would relate the markers with their respective trisomic type. The easy identification of trisomics in the segregating progeny greatly improves the efficiency of the tests. In this way 30 genes have been identified with 11 chromosomes. Several linkage groups that had previously been assumed independent were thereby associated with the same chromosome. Most of these relationships have been confirmed by standard linkage tests (Rick, Dempsey and Khush, 1964).

The remaining, unmarked chromosome, No. 12, has been tested with negative results against a total of 60 genes, most of which had been previously screened against other chromosomes. If account is taken of relative euchromatic lengths and of the previous testing of unlocated markers, 0·0045 is estimated as the probability that chromosome 12 would not have been marked in these tests. Additional evidence is thereby provided that spontaneous mutant genes are not randomly distributed among tomato chromosomes (Rick, Dempsey, and Khush, 1964). A previous report (Rick, 1959), based on standard linkage tests of spontaneous mutants, revealed a significant excess of mutants on chromosomes 2 and 11 and a tendency for markers to be concentrated in tightly linked knots.

(ii) *Secondary trisomics*

Secondaries—those trisomics in which the extra is an isochromosome—have been acquired as by-products in the progenies sired by X-rayed pollen for the purpose of inducing deficiencies (see 3 below). An example, a secondary in which the long arm of chromosome 6 is duplicated, is illustrated in plate I, fig. 1. Such trisomics can be used to place marker genes in their respective chromosome arms, although tertiary trisomics are more efficient testers and hence receive more attention in our programme.

(iii) *Tertiary trisomics*

An unexpected benefit of the induced-deficiency programme was the appearance of numerous translocation monosomics, in which the short arms of two non-homologous chromosomes had been lost after breakage in or near the centromere, their long arms having united to form a single surviving, interchange chromosome. From such monosomics have been derived the corresponding trisomic, in which the interchanged chromosome is extra—hence a tertiary trisomic. We

B

also have obtained tertiaries from reciprocal translocation heterozygotes as products of non-disjunction. To date, four different tertiary combinations have been thus obtained.

After a gene has been assigned to its proper chromosome by use of primary trisomics, it is possible to delimit it further to its respective arm by testing its segregation in the progeny of tertiary trisomics. Since in synthesising such a test the marker will usually be carried by a normal chromosome and since all haploid gametes carrying the translocated chromosome will be inviable, the segregation ratio in the diploid progeny should not be modified; but the ratio among trisomic offspring will be severely distorted if the gene is located on the arm involved in the formation of the tertiary chromosome. If the gene is situated on the arm not forming part of the extra tertiary chromosome, the ratio among the trisomic progeny should not be modified.

The induced-deficiency technique revealed that *clau* and *ful* are situated on the short arm of chromosome 4 (4S) (Khush and Rick, 1963*b*), and *e* on the long arm of chromosome 4 (4L). The same method did not, however, give information on the location of *gri* and *ven*, which were known by linkage tests to reside between *ful* and *e*. By the tertiary trisomic ratio method *gri* was located on 4L. The centromere of chromosome 4 therefore lies between *ful* and *gri* (which are only five crossover units apart).

Theoretically a set of six tertiaries with interchanges including arms of each of the tomato chromosomes would be sufficient to test the whole complement.

3. CHROMOSOME DEFICIENCIES
(i) *Segmental deficiencies*

Localisation of marker genes has been greatly expedited by use of the induced-deficiency method. This consists simply of fecundating a line homozygous for the recessive alleles, desirably also a male-sterile gene to facilitate large-scale hybridisation, with X-rayed pollen of a line homozygous for the corresponding normal alleles. Plants of mutant

---

PLATE I (*opposite*)

FIGS. 1-6. Photomicrographs of tomato pachytene chromosomes. In each pair of figures *a* is the photomicrograph, *b* is the interpretive drawing. The large, internal, faint staining body of each bivalent is the centromere. ×2,000.

FIG. 1. Trivalent association of secondary trisomic for 6L.

FIGS. 2 and 4. Interstitial deficiences for l_1. Fig. 2 Plant No. 63L746-3 and fig. 4 Plant No. 63L1012-1.

FIG. 3. Interstitial deficiency for *bu*. The presence of deficiency is indicated by a shift in the knob of the euchromatin of 8L.

FIGS. 5 and 6. Inversion of 8L. In fig. 5 the long arm is completely unpaired. In fig. 6 the inverted segment of one strand is paired non-homologously with the normal strand. The two break points of the inversion are explicitly clear in this fig. For discussion of the location of *dl* at the euchromatic breakpoint, see the text.

FIG. 7. Drawing of chromosome 8, indicating the extent of all the deficiencies and the inversion at the top of drawing and the limits within which each locus is situated are indicated at the bottom.

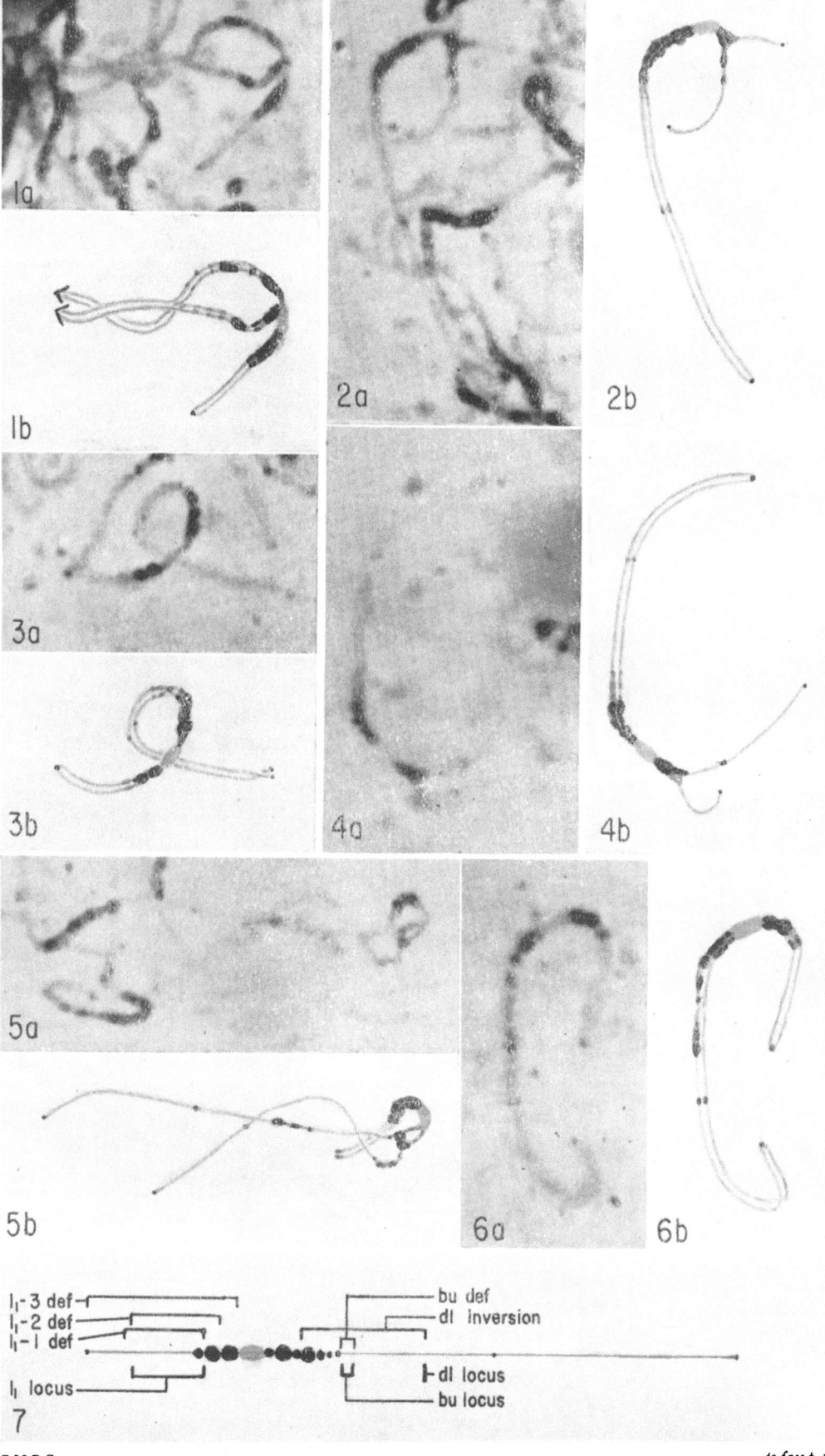

phenotype are selected in the progeny for cytological study. The combination of several markers in the female parent greatly improves efficiency of the method: additional markers on the same chromosome may mark the extent of the deficiencies, and markers on other chromosomes may detect the occurrence of accidental selfing. The test is limited to those loci for which both dominant and recessive alleles are viable and reasonably fertile in homozygous condition, and also to those that are hemizygous viable. The efficiency of the test is vastly increased if the mutant phenotype is manifest in the seedling stage.

Our first experiences (Rick and Khush, 1961) were unexpectedly rewarding. The linked genes, a_1 and hl, whose respective chromosome had not yet been determined, were selected for the test. Three terminal and one interstitial deficiency for hl delimited its locus to the proximal half of 11S, while three terminal deficiencies for a_1 proved its locus to be in the euchromatin of the long arm. Of ten plants with a_1 hl phenotype in the progeny, nine proved to be haplo-11 and one haploid.

Subsequently the method has been applied to other chromosomes, for which the following example in chromosome 8 is typical.

The tester stock selected for chromosome 8 was ms_2, dl-bu-l_1, the latter three genes having been assigned to this chromosome by trisomic and linkage tests. Pollen of a normal stock was treated with 5,000 r of X-rays and the progeny grown. In a total of 1,326 plants, two were phenotypically l_1, one bu, and one dl. The absence of any multiple mutants suggests that hemizygous deficiencies of such extent are not viable. Nine additional l_1 mutants were secured in 10,020 offspring from a similar cross in which no other marker of 8 was tested. Information is presented only for the more critical deficiencies.

Three l_1 deficiencies (one terminal and two interstitial) proved that this gene lies on 8S (plate I, figs. 2 and 4). The region common to all three embraces the terminal chromomere of the heterochromatic zone and the proximal two-fifths of the euchromatin of 8S. The single bu mutant revealed as its only cytological defect a clearly defined and consistent shift in the knob of the euchromatic of 8L (plate I, fig. 3). Since the proximal heterochromatin was intact, bu was thus delimited to the proximal two-fifths of 8L. The dl mutant proved to be heterozygous for an inversion in 8L, for which no deficiency could be detected cytologically (plate I, figs. 5 and 6). Since dl is known by crossover data to be distal to bu, the most likely site of the former is the distal break of the inversion, at a point about one-fourth of the way out on the euchromatin. It follows that bu must lie in a small section of the euchromatin adjacent to the transition and equal in extent to the bu deficiency. The extent of these deficiencies and the delimitation of loci are summarised in plate I, fig. 7. A preliminary misinterpretation of the dl inversion and of a translocation monosomic for l_1 led to our earlier erroneous conclusion that l_1 lies in 8L and bu and dl in 8S (Khush and Rick, 1963b).

Our experience with induced deficiencies permits the following generalisations concerning breakage pattern and the tolerance limits of the tomato:

(*a*) More than 95 per cent. of the X_1 mutants are visibly deficient for the locus of the tested gene. The most frequent class of deficiency is what appears to be a terminal deficiency, in which the euchromatic portion is lost, and varying amounts of the proximal heterochromatin remain. The difference cannot be detected between a simple terminal deficiency and an interstitial loss in which all euchromatin is lost and the telomere rejoined with proximal heterochromatin, although the latter would seem unlikely because it would always require breaks so close to the telomere that no euchromatin would be detected. The aforementioned translocation monosomics, in which all observed exchanges have occurred at the centromere, is another frequent category. Interstitial losses are comparatively rare, but highly diagnostic. For all classes of deficiencies about 60 per cent. of the breaks occurred in proximal heterochromatin, 25 per cent. in the centromere and 15 per cent. in euchromatic regions. Although by definition this sample is not random, the distribution of breaks corresponds roughly with that previously found in tomatoes by Barton (1954) and Gottschalk (1951).

(*b*) Whole, or nearly whole, arm losses are tolerated for either arm of chromosomes 5, 11, and 12, but only for the shorter arm of the remaining chromosomes.

(*c*) Until recently no marker had been spotted in a heterochromatic zone. In a total of 50 deficiencies for 15 genes belonging to nine chromosomes analysed to date, the only one obtained for *nv* revealed unequivocally that its locus lies in the heterochromatic zone of 9L (Khush and Rick, 1964). The configuration of chromosome 9 examined in many cells clearly revealed that the treated chromosome was deficient for all of the 9L heterochromatin except the proximal knob and the two distal ones (Khush, Rick, and Robinson, 1964). Although *nv* is thereby delimited to a heterochromatic zone, its locus is not thereby proven to be heterochromatic, because, like most proximal zones of the complement, the heterochromatic knobs are interspersed with sections that appear identical in staining capacity with euchromatin. Considering that breakage frequency is very much higher in tomato heterochromatin and that losses of heterochromatin tend to reduce viability less, the proportion of marker genes in heterochromatic zones is probably very low.

(*d*) None of the deficiencies has yet been found to be transmitted to progeny of the hemizygotes. The most extensive studies made so far have been with the a_1 and *hl* deficiencies of chromosome eleven. Five of the monosomics and 12 assorted deficiencies, including large terminals for either arm and the aforementioned interstitial in 11S were crossed with appropriate male testers of normal complement. No transmission whatever was found in 131 offspring from the monosomics and 1,429 from the deficiencies.

(e) It is likely that the primary factor limiting the extent of deficiency in X_1 is sporophytic tolerance. This impression is gained from the extreme weakness of individuals suffering the larger deficiencies. The limits (if any) tolerated by mature pollen are certainly far greater than those tolerated by pollen before complete gametogenesis, as revealed in the preceding section.

(ii) Monosomics

The appearance of monosomics as a class of deficiency deserves special consideration. As noted above, simple monosomics were found for chromosome 11. Tests repeated since the first work with that chromosome yielded more plants of haplo-11, but extensive searches, utilising markers of nine other chromosomes, have not revealed any other simple monosomics.

" Translocation monosomics "—the source of our tertiary trisomics—are, on the other hand, one of the most frequent types of deficiency encountered in our programme. This type of monosomic has a translocated chromosome but only one normal homologue of each of the two chromosomes represented in the interchange. Many different chromosomes are represented among the translocation monosomics hitherto encountered, but the missing elements so far identified have been limited to the aforementioned group of 15 shorter chromosome arms of the complement. Thus among these, as well as among the simple monosomics, the chromatin losses tolerated are highly select. Either type is highly diagnostic for the location of genes according to the method outlined for segmental deficiencies.

Hopes of using tomato monosomics in the traditional manner of monosomic analysis applied to tobacco and wheat were dashed by the failure of any monosomics to appear in the progeny of monosomics of either kind. In addition to the failure of transmission reported above for haplo-11, many hundreds of offspring from translocated monosomics have been searched without finding any. This apparent paradox of the transmission, albeit at a very low rate, of deficiencies to X_1 plants from X-rayed pollen, but the non-transmission from a deficiency heterozygote, must be explained by gametophytic elimination. Thus, whereas treated mature pollen will transmit many, perhaps all, kinds of deficiency, most of them fail to pass the full course of gametogenesis. Our past failures to obtain monosomics in the progeny of haploids, trisomics, and asynaptics are thereby explained.

4. SPECIES HYBRIDS

A remarkable degree of interspecific compatibility renders the genus *Lycopersicon* especially favourable for studies of species relationships, genic transfers, and for other purposes. Throughout this presentation *Solanum pennellii* is considered congeneric with the tomatoes, because it will hybridise with species of *Lycopersicon* and not with those of *Solanum*, because it yields partially fertile hybrids with the former (Rick, 1960), and because its chromosomes pair completely with those of *L. esculentum* and otherwise behave normally in meiosis in F_1 hybrids (Khush and Rick, 1963a). Slight chromosomal differentiation can be detected morphologically between the two species, as indicated by preferential pairing in their 4n F_1 hybrids (Rick and Smith, 1962). Even in the *bona fide* intergeneric hybrid, *L. esculentum* × *S. lycopersicoides*, the same degree of morphological similarity is found, although some bivalents tend to fall apart in later stages and the hybrid is completely sterile (Menzel, 1962: Rick, 1951). Hybrids can be produced between most combinations of the tomato species and all combinations with *L. esculentum*.

Moderate to high fertility is exhibited by all tomato species hybrids and a very high degree of homology between parental chromosomes is manifest in their regular pairing from zygotene to metaphase. Even in the widest crosses, therefore, the tomatoes contrast with certain other well investigated genera like *Gossypium*, *Nicotiana*, and *Triticum*, in which profound genomic differences have evolved. Although the former thereby lack a device that can often elucidate interspecific relationships, this disadvantage is well offset by the high degree of compatibility and hybrid fertility sufficient to permit the application of many genetic criteria of degrees of relationship.

It is in this area of experimental genetic exchange between species that tomato chromosomes can best be manipulated. Because *L. esculentum* is cross-compatible with all the other species and because it has become well known genetically, our efforts are being concentrated on its hybrids, particularly those with *L. chilense* and *S. pennellii*.

The earliest accomplishments in the interspecific chromosome engineering in tomatoes were achieved in plant breeding programmes. *Lycopersicon*, in fact, serves as an excellent example of the successful transfer of important economic traits from wild to cultivated species. Such a clearly defined monogenic trait as resistance to *Fusarium* wilt (*I*) was incorporated in *L. esculentum* from *L. pimpinellifolium* by Porte and Wellman (1941) and resistance to root-knot nematode (*Mi*) was transferred from *L. peruvianum* by Gilbert and McGuire (1955) and others.

The recent elucidation of the *esculentum* linkage groups permits far more precise studies of interspecific transfers. Thanks to the wealth of markers on most chromosomes, it is now possible to hybridise wild species with strains of *L. esculentum* in which single chromosomes are multiply marked with recessive genes. Furthermore, it is becoming feasible to specify that the characters be expressed in early seedling stages so that large numbers of progeny can be grown and scored with greatly reduced demands on space and time. As an example, a stock has recently been synthesised in which the five compatibly segregating seedling markers, *clau*, *ful*, *gri*, *e*, and *di*, have been combined to mark nearly the whole known linkage map of chromosome 4 (Khush, unpublished).

With such materials it has become possible to transfer whole chromosomes or chromosome segments from wild species to *L. esculentum* in order to study a variety of genetic phenomena. Studies of the backcrosses to *L. esculentum* from hybrids with *L. chilense*, for example, have revealed that genes suffer strong non-random segregation and that the elimination is largely, if not entirely, postsyngamic (Rick, 1963).

TABLE 1

Recombination values for chromosome 8; BC_2 to L. esculentum *from hybrid with* S. pennellii

	Total progeny	Recombinants for the intervals:			Heterogeneity x^2 (df)
		dl - bu	*bu - l_1*	*dl - l_1*	
BC_2	663	9 (1·4%)	43 (6·5%)	52 (7·8%)	4·98(4)
Control (*L. esc.*) . . .	309	not tested	56 (18·3%)	not tested	1·75(3)
Standard† . . .		7-10%	18-22%		
x^2, $BC_2 \times$ control (df) . .			29·96(1)***		

† Values compiled from many tests.

Since most of the studies on hybrids with *S. pennellii* are still in progress, only preliminary results can be presented here to exemplify some of the possibilities. A series of backcrosses has been initiated with the object of transferring certain chromosome segments from *S. pennellii* to *L. esculentum*, utilising the aforementioned recessive marker technique. By maintaining uniform environmental conditions for both the interspecific backcrosses and *esculentum* controls it has proved possible to compare recombination rates in the second backcross generation (BC_2) for marked regions of chromosomes 8 and 11. The results for chromosome 8, summarised in table 1, reveal marked reductions for both the *dl-bu* and *bu-l_1* intervals tested. The uniformity of environmental control achieved is attested by the relatively low x^2, 4·98(4), obtained for heterogeneity between the five tested lines of BC_2. On the other hand, the difference between the BC_2 lines and the control for the *bu-l_1* interval, also tested by heterogeneity x^2, is highly significant.

The data for chromosome 11 are given in table 2. Here the test for heterogeneity between BC_2 lines borders on significance, but the differences between BC_2 and controls for the two intervals tested is not significant, although backcross values are lower in both. For this chromosome it is known that a_1 and *hl* straddle the centromere, while *hl-j_1* mark the proximal part of the euchromatin in 11S. Although the populations are rather small and the data are inadequate in other respects, the contrast in results is of interest.

If these modifications of recombination are compared with the cytological features of the F_1 hybrid (Khush and Rick, 1963a), it can be seen that the much greater reduction corresponds with the substantial heteromorphy of the proximal region of chromosome 8 and the more nearly normal values with no detectable heteromorphy in chromosome 11. An effect on both tested intervals of 8 might be anticipated because the heterochromatin of the *pennellii* homologue extends well into what might be expected, on grounds of pairing, to be the *bu-dl* region.

TABLE 2

Recombination values for chromosome 11; BC_2 to L. esculentum *from hybrid with* S. pennellii

	Total progeny	Recombinants for the intervals:			Heterogeneity x^2 (df)
		$a_1 - hl$	$hl - j_1$	$a_1 - j_1$	
BC_2	782	59 (7·5%)	100 (12·8%)	152 (19·4%)	13·06* (5)
Control (*L. esc.*) . . .	117	15 (12·8%)	21 (17·9%)	36 (30·8%)	One family
Standard†		15%	20%		
x^2, $BC_2 \times$ control (df) . .		3·08 (1)	1·90 (1)		

† Values compiled from many tests.

A problem immediately raised by these results is the contradiction between these measures of reduced recombination on one hand and normal chiasma counts observed in the F_1 hybrid (Khush and Rick, 1963a) on the other. Clearly if crossing-over and chiasma formation are synonymous—a commonly accepted hypothesis—then a low value in the proximal region must be compensated for by increases distally. The compensations found in *Gossypium* chromosomes by Rhyne (1962) and Stephens (1961) provide a model for a phenomenon that might occur in the tomato hybrids. The solution of this problem must await further testing and the use of improved marker stocks.

5. DISCUSSION

Attention was attracted to the tomato by Brown's (1949) discovery of the differentiation of its pachytene chromosomes into proximal chromatic zones and distal euchromatin ones, in contrast to the well known chromosomes of *Zea*, in which no such patterning exists. Although known in a few other plant species at the time of Brown's publication, the situation he revealed was something of a novelty for chromosomes of higher plants. As the pachytene chromosomes of more angiosperm species have been investigated, however, it is becoming increasingly clear that the tomato type of differentiation is more a rule than an exception. Thus at least ten angiosperm families are represented by examples, which are listed in table 3. This pattern may actually characterise the majority of dicotyledons, while dispersed heterochromatin appears to be more widespread among monocoty-ledons; nevertheless, strongly differentiated pachytene chromosomes have also been found in the monocotyledonous genera *Agapanthus* and *Sorghum*.

In view of the widespread occurrence of differentiated chromosomes, it is essential that the cytogenetic characteristics of this type be under-stood. Differences between the two types have already been uncovered in the pattern of radiation breakage sensitivity. For reasons already presented the tomato should serve as a satisfactory model of the differentiated type for such further studies.

TABLE 3

Examples of differentiated pachytene chromosomes among genera of angiosperms

Family	Genus	Reference
Amaryllidaceae .	*Agapanthus*	Belling, 1928
Gramineae . .	*Sorghum*	Magoon *et al.*, 1961
Ranunculaceae .	*Aquilegia*	Linnert, 1961
Cruciferae . .	*Brassica*	Robbelen, 1960
Balsaminaceae . .	*Impatiens*	Smith, 1934
Onagraceae . .	*Oenothera*	Marquardt, 1937
Labiatae . .	*Salvia*	Linnert, 1955
Solanaceae . .	*Lycopersicon*	Brown, 1949
	Atropa, Datura, Hyoscyamus, Nicandra, Physalis, Saracha, Schizanthus, Scopolia, Solanum, Withania.	Gottschalk, 1954*b*
Gesneriaceae	*Aeschynanthus, Alloplectus, Columnea, Iso-loma, Klugia, Sinningia.*	Eberle, 1956
Compositae	*Layia, Tragopogon.*	Scherz, 1957

A question that logically arises concerning the high chromosome number of *Lycopersicon* (and most other genera of the Solanaceae) is, what role, if any, has polyploidy played in its evolution. Rick and Barton (1954) faced this problem and, on the basis of low tolerance of extra chromosomes and other evidence, concluded that the tomato shows no evidence of recent duplication within its genome. Gottschalk (1956) arrived at the opposite conclusion; specifically, that the basic number of the genus is six chromosomes. His argument is based chiefly on presumed resemblances between certain chromosomes of the haploid set, and evidences of secondary pairing. In arriving at his conclusion he places little emphasis on intolerance of aneuploidy, complete lack of genetic evidence for duplication and the low level of chromosome pairing in haploid tomatoes. In fact, the weighting of evidence to support his thesis is astonishing. Thus, in regard to the lack of bivalents in diakinesis of haploid tomatoes, he states (p. 70): " Die Befunde am Pachytän können somit durchaus als Hinweis für die Annahme aufgefasst werden, dass die Grundzahl von *L. esculentum* $x = 6$ ist. Das Fehlen der Bivalentenbildung in der Diakinese haploider Tomaten ist kein Gegenbeweis für diese Annahme. Leider wurde das Pachytän haploider *Lycopersicon*-Arten noch nicht eingehend bearbeitet." It is thereby implied that pairing at pachytene is more indicative of homologies than pairing at later stages, yet most workers have arrived at the opposite conclusion. Thus McClintock (1933) showed convincingly that chromosome pairing at pachytene in *Zea* can take place irrespective of homology. In her studies of chromosome pairing in species hybrids of *Gossypium*, Brown (1954) concludes: " Pachytene pairing in sterile as well as fertile hybrids appears as intimate as within species. Pachytene pairing in hybrids cannot be considered a measure of chromosome or species differentiation, and metaphase pairing remains the best criterion of chromosome homology as now defined." It has been our experience also (Khush, Rick and Robinson, 1964; and unpublished) that a monosomic region tends to pair in various ways within itself or with other monosomic regions apparently without regard to homology. Also see fig. 6 in this paper for non-homologous pairing.

The great weight placed by Gottschalk on secondary pairing is also open to question. Gröber (1961) has dealt with this problem in the tomato in detail and concludes that, since the association of bivalents in later meiotic stages results from earlier, non-specific attractions between telomeres and between centromeres, they cannot be considered proof of homologies. In our own research, embracing a survey of thousands of tomato PMC's, we have not encountered any instances of attachment between bivalents in diploid tomatoes.

A closer examination of the presumed similarities between different chromosomes of the tomato set is also merited. In an earlier contribution, Gottschalk (1954a) suggests that the following " homologous " pairs can be identified: 3-4, 7-8, 9-10, 5-12 [numbered according to Barton's (1950) classification]; yet scrutiny of any of these pairs reveals remarkable differences between them. The constant features of the heterochromatic elements in chromosomes 3 and 4, for example, differ to a great extent. The classification of 5 with 12 raises further doubts, because in respect to arm ratios they resemble each other no more than either one resembles 11. Since such classifications are often subjective, they carry little weight as evidence for duplication of chromatin in phylogeny.

In such comparisons genetic evidence is often more cogent. Duplications of whole chromosomes or chromosomal segments should lead to duplication of loci and groups of linked genes. Yet no convincing case of duplicate inheritance has been reported for tomatoes, and the known linkage groups for the aforementioned " homologues " do not reveal the slightest similarity in genic content, order, or distance. Another consequence logically expected from duplication would be similarity in phenotype for the corresponding primary trisomics. Yet more contrasting pairs could scarcely be selected than triplo-3-4, 7-8, 9-10, and 5-12. What slight resemblances can be detected between the exceedingly diverse array of trisomic phenotypes are to be found between triplo-7 and 10, 3 and 6, 4 and 11.

Finally we would like to emphasise again the remarkably low tolerance of aneuploidy and structural deficiencies. Monosomy, now extensively tested for 11 chromosomes, has been found only for number 11. The loss of only certain chromosome arms, furthermore, is tolerated either singly or in pairs in the translocation monosomics. Such are the sporophytic tolerances; the gametophytic limits are extremely narrow. Thus far we have detected no transmission through either pollen or egg cells of any deficiency, no matter what size. Certainly this evidence provides very little support for the duplication hypothesis.

Acknowledgments.—It is a pleasure to acknowledge support of parts of this research by grants G.M. 06209 of the National Institutes of Health, U.S.P.H.S. and G.B. 1902 of the National Science Foundation.

6. REFERENCES

BARTON, D. W. 1950. Pachytene morphology of the tomato chromosome complement. *Amer. J. Bot.*, *37*, 639-643.

BARTON, D. W. 1954. Comparative effects of X-ray and ultra-violet radiation on the differentiated chromosomes of the tomato. *Cytologia*, *19*, 157-175.

BELLING, J. 1928. Contraction of chromosomes during maturation divisions in *Lilium* and other plants. *Univ. Calif. Publ. Bot.*, *14*, 335-343.

BROWN, M. S. 1954. A comparison of pachytene and metaphase pairing in species hybrids of *Gossypium*. *Genetics*, *39*, 962-963.

BROWN, S. W. 1949. The structure and meiotic behaviour of the differentiated chromosomes of tomato. *Genetics*, *34*, 437-461.

EBERLE, P. 1956. Cytologische Untersuchungen an Gesneriaceen. I. Die Struktur der Pachytänchromosomen, sowie eine Reihe neu bestimmer Chromosomenzahlen. *Chromosoma*, *8*, 285-316.

GILBERT, J. C., AND MCGUIRE, D. C. 1956. Inheritance of resistance to severe root knot from *Meloidogyne incognita* in commercial-type tomatoes. *Proc. Amer. Soc. Hort. Sci.*, *68*, 437-442.

GOTTSCHALK, W. 1951. Der Vergleich von röntgen- und chemisch induzierten Chromosomenmutationen in Pachytän von *Solanum lycopersicum*. *Chromosoma*, *4*, 342-358.

GOTTSCHALK, W. 1954a. Die Grundzahl der Gattung *Solanum* und einiger *Nicotiana*-Arten. *Ber. Deuts. Bot. Ges.*, *67*, 369-376.

GOTTSCHALK, W. 1954b. Die Chromosomenstruktur der Solanaceen unter Berücksichtigung phylogenetischer Frägestellungen. *Chromosoma*, *6*, 539-626.

GOTTSCHALK, W. 1956. Die Cytologie der Kulturtomate und ihrer wildwachsenden Verwandten. *Biblio. Genetica*, *17*, 1-109.

GRÖBER, K. 1961. Multivalente Assoziationen in diploiden Pollenmutterzellen der Tomate (*Lycopersicon esculentum* Mill.) und ihre Bedeutung für das Problem der Chromosomengrundzahl. *Kulturpfl.*, *9*, 146-162.

KHUSH, G. S., AND RICK, C. M. 1963a. Meiosis in hybrids between *Lycospersicon esculentum* and *Solanum pennellii*. *Genetica*, *33*, 167-183.

KHUSH, G. S., AND RICK, C. M. 1963b. Localizing genes by means of induced chromosomal deficiencies in tomatoes. *Proc. 11th Int. Congr. Gen.*, *1*, 117.

KHUSH, G. S., AND RICK, C. M. 1964. Heterochromatic vs. euchromatic deficiencies of marker genes of the tomato. *Genetics*, *50*, 264.

KHUSH, G. S., RICK, C. M., AND ROBINSON, R. W. 1964. Genetic activity in a heterochromatic chromosome segment of the tomato. *Science*, *145*. (In press.)

LESLEY, J. W. 1928. A cytological and genetical study of progenies of triploid tomatoes. *Genetics*, *13*, 1-43.

LESLEY, J. W.   1932.   Trisomic types of the tomato and their relations to the genes. *Genetics, 17,* 545-559.

LINNERT, G.   1955.   Die Struktur der Pachytänchromosomen in Euchromatin und Heterochromatin und ihre Auswirkung auf die Chiasmabildung bei *Salvia*-Arten. *Chromosoma, 7,* 90-128.

LINNERT, G.   1961.   Cytologische Untersuchungen an Arten und Artbastarden von *Aquilegia*. II. Die Variabilität der Pachytänchromosomen. *Chromosoma, 12,* 585-606.

MAGOON, M. L., SHAMBULINGAPPA, K. G., AND RAMANNA, M. S.   1961.   Chromosome morphology and meiosis in some Eu-sorghums. *Cytologia, 26,* 236-252.

MARQUARDT, H.   1937.   Die Meiosis von Oenothera. I. *Zeits. Zellf. Mik. Anat., 27,* 159-210.

MCCLINTOCK, B.   1933.   The association of non-homologous parts of chromosomes in mid-prophase of *Zea mays*. *Zellf. Mik. Anat., 19,* 191-237.

MENZEL, M. Y.   1962.   Pachytene chromosomes of the intergeneric hybrid *Lycopersicon esculentum* × *Solanum lycopersicoides*. *Amer. J. Bot., 49,* 605-615.

PORTE, W. S., AND WELLMAN, F. L.   1941.   Development of interspecific tomato hybrids of horticultural value and highly resistant to *Fusarium* wilt. *U.S. Dept. Agr. Circ., 584,* 1-19.

RHYNE, C. L.   1962.   Enhancing linkage-block breakup following interspecific hybridization and backcross transference of genes in *Gossypium hirsutum* L. *Genetics, 47,* 61-69.

RICK, C. M.   1959.   Non-random gene distribution among tomato chromosomes. *Proc. Nat. Acad. Sci.* (U.S.A.), *45,* 1515-1519.

RICK, C. M.   1960.   Hybridization between *Lycopersicon esculentum* and *Solanum pennellii*: phylogenetic and cytogenetic significance. *Proc. Nat. Acad. Sci.* (U.S.A.), *46,* 78-82.

RICK, C. M.   1963.   Differential zygotic lethality in a tomato species hybrid. *Genetics, 48,* 1497-1507.

RICK, C. M., AND BARTON, D. W.   1954.   Cytological and genetical identification of the primary trisomics of the tomato. *Genetics, 39,* 640-666.

RICK, C. M., DEMPSEY, W. H., AND KHUSH, G. S.   1964.   Further studies on the primary trisomics of the tomato. *Canad. J. Genet. Cytol., 5,* 93-108.

RICK, C. M., AND KHUSH, G. S.   1961.   X-ray-induced deficiencies of chromosome 11 in the tomato. *Genetics, 46,* 1389-1393.

RICK, C. M., AND KHUSH, G. S.   1962.   Preferential pairing in tetraploid tomato species hybrids. *Genetics, 47,* 979-980.

RICK, C. M., AND NOTANI, N. K.   1961.   The tolerance of extra chromosomes by primitive tomatoes. *Genetics, 46,* 1231-1235.

ROBBELEN, G.   1960.   Beiträge zur Analyse des *Brassica*-Genoms. *Chromosoma, 11,* 205-228.

SCHERZ, C.   1957.   Die Chromosomenstruktur in der meiotischen Prophase einiger Compositen. *Chromosoma, 8,* 447-457.

SMITH, F. H.   1934.   Prochromosomes and chromosome structure in *Impatiens*. *Proc. Amer. phil. Soc., 74,* 193-215.

SOOST, R. K.   1958.   Progenies from sesquidiploid F$_1$ hybrids of *Lycopersicon esculentum* and *L. peruvianum*. *J. Hered., 49,* 208-213.

STEPHENS, S. G.   1961.   Recombination between supposedly homologous chromosomes of *Gossypium barbadense* L. and *G. hirsutum* L. *Genetics, 46,* 1483-1500.

HAPLOIDY AS A NEW APPROACH TO THE CYTOGENETICS AND BREEDING OF *SOLANUM TUBEROSUM**

S. J. PELOQUIN, R. W. HOUGAS and A. C. GABERT

Departments of Genetics and Horticulture, University of Wisconsin, Madison, U.S.A.

1. INTRODUCTION

HAPLOIDS (2n = 24) of *Solanum tuberosum* L. (2n = 48) are unique material for cytogenetic and breeding studies of this species. They offer the advantages of (*a*) disomic rather than tetrasomic inheritance patterns, and (*b*) a new direct approach to gene transfer from the numerous wild and cultivated, tuber-bearing, 24-chromosome *Solanum* species. They also provide new opportunities for the investigation of many genetic, cytological and evolutionary problems in the tuberous Solanums (Hougas and Peloquin, 1958).

This paper is concerned with the problems related to obtaining haploids and the progress made using haploids in genetic, cytogenetic and breeding researches.

2. RESULTS

(i) *Initial attempt to obtain haploids*

Twenty-eight haploid plants, representing seven selections of the common potato, were obtained in the original trials. The haploids were found among seedling progenies of interploid matings, Tuberosum (2n = 48) × Phureja (2n = 24), designed to make detection of haploids relatively easy. The majority of the haploids were thrifty plants, 22 flowered, 21 were female fertile, and two of the 29 were also male fertile (Hougas, Peloquin and Ross, 1958; Peloquin and Hougas, 1960).

The success of the initial attempt to isolate haploids was encouraging both in terms of the number of haploids obtained and in their fertility. It was evident, however, that if the potential of the haploid approach was to be fully explored, sizeable stock-piles of haploids, derived from a wide range of selections of Tuberosum and Andigena, would be essential. Consequently, investigations directed at finding methods of increasing haploid frequency were initiated.

(ii) *Increasing haploid frequency*

The results of researches related to increasing haploid frequency have been most rewarding. They are summarised briefly in the following paragraphs.

* Supported in part by research grants from the National Science Foundation and the Rockefeller Foundation.
The epithet *Solanum tuberosum* is used in the sense of Dodds (1962).

Decapitation of the seed-parent has resulted in a 10-15-fold increase in fruits per pollination and a consequent increase in haploid frequency. This technique (Peloquin and Hougas, 1959) consists of decapitating the upper portion of the plant at the time the first flower opens and placing the "decapitant" in a water-filled container in an air-conditioned greenhouse (plate I, fig. 1). Decapitation has the added advantages of being easily used under various controlled environments and readily adaptable to large scale operations (more than 100,000 pollinations were made at the Potato Introduction Station, Sturgeon Bay, Wisconsin, each summer from 1959 through 1962). It should also be pointed out that it is considerably easier to pollinate at table height in an air-conditioned greenhouse than it is in the field. Employment of this technique has been of considerable help in testing the effect of seed-parent and "pollinator" (pollen source) on haploid frequency.

Selection of the seed-parent can result in about a 10-fold increase in haploid frequency. The seed-parents Merrimack and Wis. Ag. 231, superior in regard to haploid frequency, average more than 10 haploids per 100 fruit. This is in contrast to an overall frequency of one haploid per 100 fruit in the other seed-parents tested (Hougas, Peloquin and Gabert, 1964; Gabert, 1963).

Most important and somewhat surprising was the finding that the "pollinator" has a very significant effect on haploid frequency. "Pollinators" 1·1, 1·3, and 1·22, selections of Phureja (P. I. 225682), were superior. Haploid frequencies of approximately 10 per 100 fruit were obtained when these "pollinators" were used with a wide range of seed-parents; 85 other pollinators tested averaged just over one haploid per 100 fruit with the same seed-parents. The use of superior "pollinators" resulted in increased haploid frequencies, in the range of 5-15-fold, from almost all of 82 seed-parents tested (Hougas, Peloquin, and Gabert, 1964; Gabert, 1963).

Other factors tested for their effect on haploid frequency were (1) temperature, (2) growing decapitants in a nutrient solution, (3) multiple pollinations, and (4) delayed pollinations. Constant (60°F., 75°F., and 85°F.), artificially alternating, and naturally changing temperatures in an air-conditioned greenhouse (55°-60°F. at night and 70°-75°F. in

---

PLATE I (*opposite*)

FIG. 1. Decapitants of *S. tuberosum* in a greenhouse.

FIG. 2. 24-chromosome F_1 hybrids (Tuberosum haploids × *S. simplicifolium*).

FIG. 3. Pachytene in an Andigena haploid.

FIG. 4. Heteromorphic short arm of nucleolus-associated chromosome in an Andigena haploid.

FIG. 5. Variability in tuber characters of 24-chromosome Phureja-haploid hybrids.

FIG. 6. Tubers from single hills of parents and F_1 hybrid—left to right; Phureja, F_1 hybrid, haploid.

FIG. 7. Tubers from 24-chromosome F_1 hybrids (Phureja × haploid).

FIG. 8. Tubers from BC [haploid × (Phureja × haploid)].

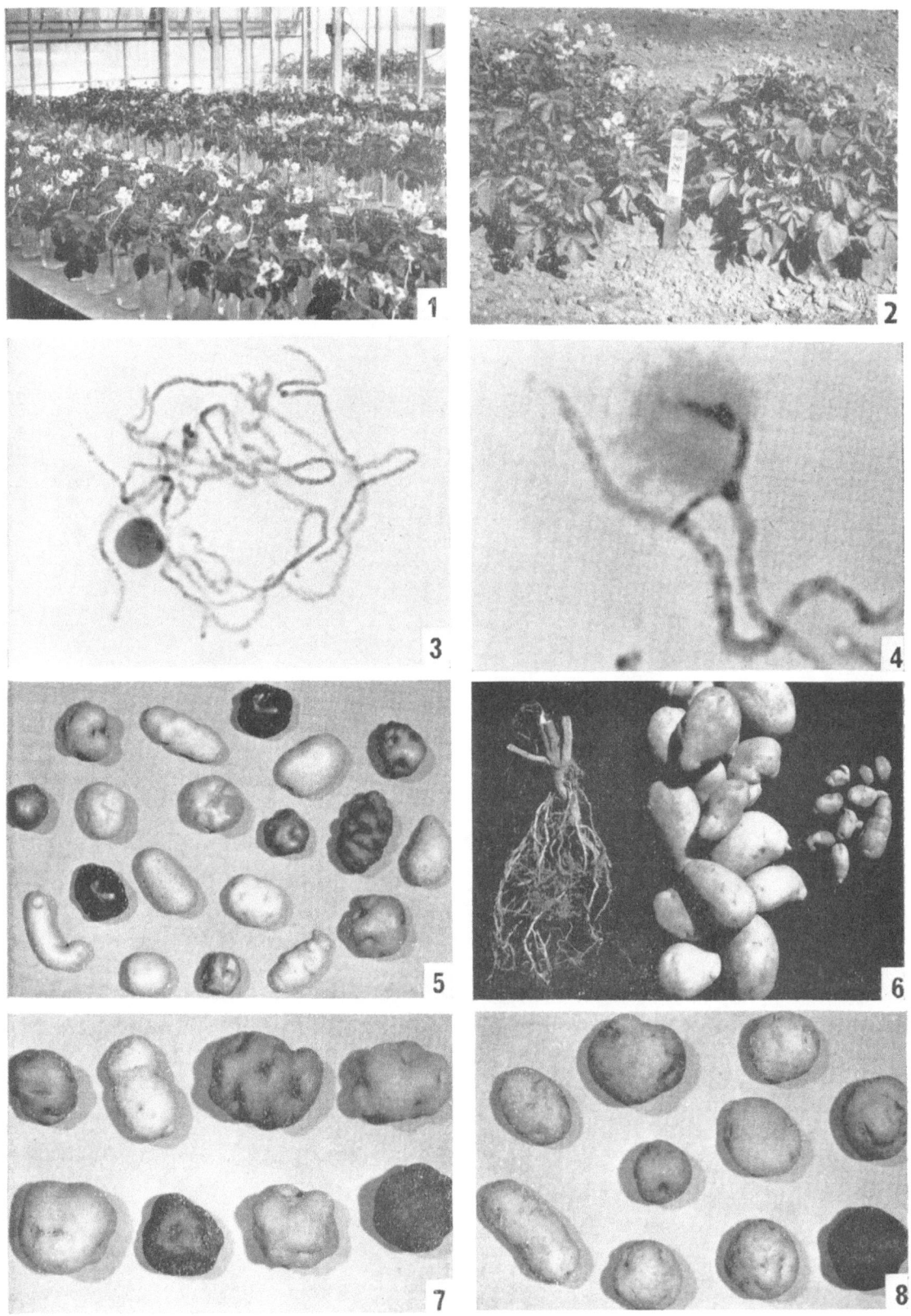

the daytime) were compared. The latter conditions gave the best results (Gabert, 1963). Wöhrmann (1964) also found that temperature did not significantly affect haploid frequency. Results with (2), (3) and (4) were all negative.

Haploid frequencies in the range of 35-80 per 100 fruit have been obtained from combinations of elite pollinators and elite seed-parents using the decapitation technique. Up to the present, more than 6300 haploids representing 23 commercial varieties, 45 breeding stocks, and 14 Andigena selections have been obtained as a result of utilising the improved materials and techniques (Gabert, Peloquin, and Hougas, 1964). Our results together with those from other laboratories (Bender, 1963; Baerecke, Frandsen, and Ross, personal communication; Dionne, personal communication; Kawakami and Matsubayashi, 1960; Rothacker and Schäfer, 1961; and Wöhrmann, 1963) indicate that obtaining haploids will not be difficult.

(iii) Heritability and nature of "pollinator" effect

High haploid frequencies have made it possible to study heritability and nature of the "pollinator" effect. Heritability studies were undertaken with the knowledge that the variation between superior and inferior "pollinators" was discontinuous and that superior "pollinators" occurred infrequently. Results of testing F_1 families from superior × superior, inferior × inferior, and superior × inferior matings clearly demonstrate that superior influence on haploid frequency is recessive. Superior "pollinators" were obtained only from superior × superior matings. Results from F_2 and BC_1 generations indicate that the "pollinator" effect is qualitatively inherited with either one or few genes involved (Gabert, Hougas, and Peloquin, 1963).

The "pollinator" appears to operate through the endosperm in affecting haploid frequency. The clue to this hypothesis came from cytological observations of developing seeds following Tuberosum (2n = 48) × Phureja (2n = 24) matings. It was found that haploid (2n = 24) embryos were regularly associated with hexaploid rather than the expected pentaploid endosperms, and that pentaploid endosperms regularly stopped developing at an early stage (Wangenheim, Peloquin, and Hougas, 1960; Bender, 1963).

Two possible ways of obtaining hexaploid endosperms are (1) gametes with 24 chromosomes function, one going to the central cell of the female gametophyte, (2) gametes with 12 chromosomes function, both going to the central cell of the female gametophyte. Two lines of indirect evidence support (2). One is the observation that one-third of the developing seeds with hexaploid endosperms did not possess embryos. It is difficult to imagine fertilisation of the egg often failing thus, if a male gamete is available for fertilisation. The second evidence is that more haploids (2n = 24) than tetraploids (2n = 48) were found in the progeny of certain Tuberosum (2n = 48) × Phureja (2n = 24) matings. This appears improbable under (1), since, if the functioning of 24-chromosome gametes is necessary for haploids to develop, more tetraploids than haploids would be expected in the progeny of Tuberosum × Phureja matings. A superior pollinator appears, therefore, to be one that contributes two 12-chromosome gametes to the endosperm leaving none available to fertilise the egg (Peloquin, Hougas, and Gabert, 1963).

(iv) Genetics

The value of the haploids in genetic studies is essentially dependent on the fertility of the haploids, their crossability with 24-chromosome *Solanum* species, and the vigour, fertility, and variability present in the species-haploid hybrids and their derivatives. A workable level of fertility is present among the haploids. In any one growing season approximately 50 per cent. of the haploids flowered, 30 per cent. were female fertile, and 3 per cent. were male fertile. Since both flowering and fertility in Tuberosum are considerably influenced by environment, the actual frequency of fertile haploids is probably considerably higher than that found in any one season. Crosses between Tuberosum haploids and between Tuberosum and Andigena haploids have been successful. Although these hybrids are somewhat difficult to obtain, the results are worth the extra effort, since many F_1 hybrids exceed their parents in both vigour and fertility. The high incidence of male fertility is particularly promising, since it represents one possible alternative means of obtaining increased male fertility in haploids.

Haploids have been successfully crossed with 23 of the 24-chromosome, tuber-bearing species representing five taxonomic series (Hougas and Peloquin, 1960). Similar results have been obtained by Bender (1963) with 20 species, and Matsubayashi (1961) with two species. The majority of the F_1 hybrids were vigorous and flowered profusely. Two F_1 hybrids (haploid × *S. simplicifolium* Bitt.) are shown in plate I, fig. 2.

The fertility of F_1 hybrids between haploids and diploid species is generally high. Male sterility is, however, present in certain hybrids and is conditioned by the direction of the cross as well as the haploid and species involved. When the cultivated diploids (Phureja and Stenotomum) are used as staminate parents in crosses with haploids, the F_1 hybrids are male sterile. Hybrids obtained from the reciprocal cross are, however, male fertile (Ross, Peloquin and Hougas, 1964). Furthermore, hybrids between haploids and the wild species *S. simplicifolium* Bitt., *S. neohawkesii* Ochoa, and *S. chacoense* Bitt., are male fertile regardless of the direction the cross is made.

Hybrids between haploids and the cultivated diploids Phureja and Stenotomum appear particularly promising for genetic studies. Besides possessing good vigour and fertility, these hybrids tuberise relatively well under long-day conditions and are extremely heterogeneous. The variation in tuber characters (shape, depth of eyes, number of eyes, height of eyebrow) present in this material is illustrated in plate I, fig. 5. This same variation in other characters (tuber dormancy, specific gravity, chipping quality, and resistance to scab) is also present in these hybrids and their derivatives. The recent finding of both a high incidence and level of self-compatibility among particular families of normally self-incompatible Phureja-haploid hybrids (Cipar, 1964; Smiley, 1963) is an additional reason in favour of their use in genetic studies.

Haploids are valuable for investigating many aspects of self-incompatibility in Solanums. For example, the fact that haploids which are male and female fertile are self-incompatible is evidence for the presence of an S allele system in tetraploid potatoes. Further, S allele analysis of haploids by use of Phureja-haploid F_1 hybrids indicates that several haploids are S allele homozygotes. These have been useful tools for determining S alleles present in other clones and for synthesising a series of hybrid testers, $S_{1.2}-S_{1.n}$, (Cipar, Peloquin and Hougas, 1964). In exploring S allele relations in groups Tuberosum, Andigena, and Phureja, two S alleles were found in Tuberosum haploids which appear identical to S alleles found in Andigena haploids and Phureja. The S locus appears, therefore, to be a useful tool for evolutionary studies in the tuber-bearing Solanums.

(v) Cytogenetics

Haploids provide a simpler and more direct approach to several cytogenetic problems. Meiosis was studied in haploids ($2n = 24$) of Tuberosum ($2n = 48$) and Andigena ($2n = 48$), and in inter-haploid hybrids within and between these groups (Yeh, Peloquin and Hougas, 1964). Twelve bivalents were regularly present at diakinesis and metaphase I in the majority of clones analysed. Univalents, delayed separation of bivalents, and incomplete divisions were the most frequent irregularities in the remaining clones. The observation that meiosis was more regular in most interhaploid hybrids than in their parents, plus additional evidence from haploid-species hybrids, indicates that the high frequency of meiotic irregularities in a few clones may be the result of the effect of the genotype on the meiotic process.

Recent success with pachytene analysis also indicates regular chromosome pairing in haploids and interhaploid hybrids. Plate I, fig. 3 illustrates the chromosome pairing at pachytene in an Andigena haploid. Although it is a rare sporocyte in which all 12 chromosomes can be followed with certainty, individual chromosomes are often separate and thereby available for analysis. Through a combined study of a few entire sporocytes and individual chromosomes it has been possible to identify each of 12 chromosomes at pachytene (Yeh

c

and Peloquin, 1965). The only gross structural difference detected is in the heterochromatic region of the short arm of the nucleolus-associated chromosome. Chromosomes vary in amount of heterochromatin distal to the nucleolus-associated constriction (plate I, fig. 4). Haploids from the same parent were both heteromorphic and homomorphic for this region.

The results from pachytene and metaphase I both support the concept that the two genomes of Andigena are similar to each other and to the genomes of Tuberosum. Evidence from fertility, crossability, and meiotic studies of haploids and their derivatives all indicates that there are no significant barriers to the free exchange of genetic material between the genomes. It seems logical, therefore, to consider both Andigena and Tuberosum essentially autotetraploids.

(vi) Breeding

A very unexpected and pleasant surprise was the tuber yield of many Phureja-haploid F_1 hybrids. These hybrids substantially outyielded their parents (plate I, fig. 6). Furthermore, the yield of several selections from these 24-chromosome F_1 populations equalled the yield of commercial tetraploid varieties. The majority of the F_1 hybrids (plate I, fig. 7) have undesirable characteristics such as deep eyes, short dormancy, and high tuber sets. However, clones with shallow eyes, long dormancy and low tuber set were found in small BC [haploid × (Phureja × haploid)] families (plate I, fig. 8). The vigour, yield, fertility, and variability of these 24-chromosome materials suggest it is worth exploring the breeding of potatoes at the diploid level.

The use of haploids in potato breeding should not, of course, be restricted to breeding at the diploid level. Other approaches would include: (1) crossing standard varieties with selected 24-chromosome clones possessing desirable characteristics such as resistance to disease. Since the progeny of this 4x-2x cross are mainly tetraploids, it provides a method of incorporating desirable genes from the 24-chromosome material while maintaining good horticultural characteristics; (2) synthesis of triploids as an initial step in exploring the effect of different levels of ploidy on vigour and productivity. The problem at present with this approach is the difficulty of obtaining triploids in large numbers; (3) employing the hypothetical analytical breeding scheme as outlined by Chase (1963). This essentially consists of intensive breeding and selection at the diploid level followed by resynthesis of the tetraploid. Testing this hypothetical scheme appears reasonable, since all the operations outlined have been accomplished in separate lines of research. The utilisation of haploids in potato breeding would appear to merit considerable attention in view of the large amount of effort presently directed to potato breeding at the tetraploid level.

3. REFERENCES

BENDER, K. 1963. Über die Erzeugung und Entstehung dihaploider Pflanzen bei *Solanum tuberosum*. *Z. Pflanzenzücht.*, *50*, 141-166.

CHASE, S. S. 1963. Analytic breeding in *Solanum tuberosum* L.—a scheme utilizing parthenotes and other diploid stocks. *Can. J. Genet. Cytol.*, *5*, 359-363.

CIPAR, M. S. 1964. Self-compatibility in hybrids between Phureja and haploid Andigena clones of *Solanum tuberosum*. *Europ. Potato J.*, *7*, 152-160.

CIPAR, M. S., PELOQUIN, S. J., AND HOUGAS, R. W. 1964. Inheritance of incompatibility in hybrids between *Solanum tuberosum* haploids and diploid species. *Euphytica*, *13*, 163-172.

DODDS, K. 1962. Classification of cultivated potatoes. In: *The Potato and Its Wild Relatives*, by D. S. Correll, Texas Research Foundation, Renner, Texas, 517-539.

GABERT, A. C. 1963. Factors influencing the frequency of haploids in the common potato (*Solanum tuberosum* L.). Ph.D. Thesis, University of Wisconsin.

GABERT, A. C., HOUGAS, R. W., AND PELOQUIN, S. J. 1963. Heritability of the pollinator effect on haploid frequency in the common potato, *Solanum tuberosum* L. *Agronomy Abstracts*, p. 80.

GABERT, A. C., PELOQUIN, S. J., AND HOUGAS, R. W. 1964. Haploids of *Solanum tuberosum* L. isolated from 123 seed-parents. *Agronomy Abstracts*, p. 67.

HOUGAS, R. W., AND PELOQUIN, S. J. 1958. The potential of potato haploids in breeding and genetic research. *Amer. Potato J.*, *35*, 701-707.

HOUGAS, R. W., AND PELOQUIN, S. J. 1960. Crossability of *Solanum tuberosum* haploids with diploid *Solanum* species. *Europ. Potato J.*, *3*, 325-330.

HOUGAS, R. W., PELOQUIN, S. J., AND GABERT, A. C. 1964. Effect of seed-parent and pollinator on frequency of haploids in *Solanum tuberosum*. *Crop Sci.*, *4*, 593-595.

HOUGAS, R. W., PELOQUIN, S. J., AND ROSS, R. W. 1958. Haploids of the common potato. *J. Hered.*, *47*, 103-107.

KAWAKAMI, K., AND MATSUBAYASHI, M. 1960. Studies on the haploid plants of *Solanum tuberosum*. I. Morphological characteristics of the polyhaploid plants. *Jap. J. Breed.*, *10*, 9-18.

MATSUBAYASHI, M. 1961. Cytogenetic studies in *Solanum*. Section Tuberarium, with special reference to the interspecific relationships. Diss. Kyoto University.

PELOQUIN, S. J., AND HOUGAS, R. W. 1959. Decapitation and genetic markers as related to haploidy in *Solanum tuberosum*. *Europ. Potato J.*, *2*, 176-183.

PELOQUIN, S. J., AND HOUGAS, R. W. 1960. Genetic variation among haploids of the common potato. *Amer. Potato J.*, *37*, 289-297.

PELOQUIN, S. J., HOUGAS, R. W., AND GABERT, A. C. 1963. Haploid frequency in *Solanum tuberosum* following 4x-2x matings: Nature of the " pollinator " effect. *Agronomy Abstracts*, p. 87.

ROSS, R. W., PELOQUIN, S. J., AND HOUGAS, R. W. 1964. Fertility of hybrids from *Solanum phureja* and haploid *S. tuberosum* matings. *Europ. Potato J.*, *7*, 81-89.

ROTHACKER, D., AND SCHÄFER, G. 1961. Einige Untersuchungen über haploide Pflanzen von *Solanum tuberosum*. *Züchter*, *31*, 289-297.

SMILEY, J. H. 1963. Behaviour of self-fertility and its utilization in diploid *Solanum* species-haploid *S. tuberosum* hybrids. Ph.D. Thesis, University of Wisconsin.

WANGENHEIM, K-H. V., PELOQUIN, S. J., AND HOUGAS, R. W. 1960. Embryological investigations on the formation of haploids in the potato (*Solanum tuberosum*). *Z. Vererbungslehre*, *91*, 391-399.

WÖHRMANN, K. 1963. Über den Einfluss von röntgenbestrahltem Pollen auf den Anteil dihaploider Sämlinge in den Kreuzungnachkommenschaften von *Solanum tuberosum* L. × *Solanum phureja* Juz. et Buk. *Z. Pflanzenz.*, *50*, 132-139.

WÖHRMANN, K. 1964. Über den Einfluss der Temperatur auf die Dihaploidenrate bei *Solanum tuberosum* L. *Z. Pflanzenzücht.*, *52*, 1-7.

YEH, B. P., AND PELOQUIN, S. J. 1965. Pachytene chromosomes of the potato (*Solanum tuberosum* Group Andigena.) *Amer. J. Bot.* (In press.)

YEH, B. P., PELOQUIN, S. J., AND HOUGAS, R. W. 1964. Meiosis in *Solanum tuberosum* haploids and haploid-haploid F$_1$ hybrids. *Can. J. Genet. Cytol.*, *6*, 393-402.

NULLISOMIC-TETRASOMIC COMBINATIONS IN HEXAPLOID WHEAT*

E. R. SEARS†

U.S. Department of Agriculture and Field Crops Department,
University of Missouri, Columbia, Missouri, U.S.A.

1. INTRODUCTION

THE discovery that the 21 different chromosomes of common wheat (*Triticum aestivum* L.) fall into seven homoeologous groups of three (Sears, 1952, 1954) was based primarily on the ability of each tetrasome to compensate for the nullisome of each of the other two chromosomes of the same group. Supporting evidence has come from the finding of Okamoto and Sears (1962) that the pairing in haploids is largely between chromosomes belonging to the same homoeologous group, and from the work of Riley and Kempanna (1963), who found only pairing of homoeologues when increased pairing was induced by the absence of chromosome 5B.

Some brief general observations were made by Sears and Okamoto (1956) concerning the 40 within-group (compensating) and 47 between-group (non-compensating) combinations then in existence. Those observations will be extended in the present paper to include the additional two compensating and 14 non-compensating combinations now available. Detailed data on origins will also be presented, along with figures on fertility and information on morphological characters.

The system of numbering developed by Sears (1958) and Okamoto (1962) will be employed. These numbers correspond as follows to those previously used: 1A = XIV, 1B = I, 1D = XVII, 2A = XIII, 2B = II, 2D = XX, 3A = XII, 3B = III, 3D = XVI, 4A = IV, 4B = VIII, 4D = XV, 5A = IX, 5B = V, 5D = XVIII, 6A = VI, 6B = X, 6D = XIX, 7A = XI, 7B = VII, 7D = XXI.

The terms nullisomic, monosomic, etc., will usually be shortened to nulli, mono, etc. Where a pair of chromosome designations (as 1A-1B) are given, the chromosomes concerned are of reduced and increased dosage, respectively. A single designation (as tetra-3D) implies that the other 20 chromosomes are disomic.

One of the nulli-tetras (5D-5B) was kindly provided by Dr. Ralph Riley of the Plant Breeding Institute, Cambridge. All the materials belonged to the variety Chinese Spring.

* Co-operative investigations of the Crops Research Division, Agricultural Research Service, U.S. Department of Agriculture, and the Field Crops Department, Missouri Agricultural Experiment Station. Journal Series No. 2796 of the Missouri Station. Approved by the Director.

† Research Geneticist, Crops Research Division, A.R.S., U.S.D.A. Address: Curtis Hall, University of Missouri.

2. COMPENSATING COMBINATIONS

(i) *Origin*

Of the 42 nullisomic-tetrasomics involving chromosomes of the same homoeologous group, 25 were synthesised by pollinating a mono-somic by a tetrasomic, backcrossing the resulting mono-tri by the tetrasomic, and selfing the resulting mono-tetra. The other 17 combinations arose in various ways (fig. 1), five spontaneously (without crossing).

In order to obtain the mono-tetras from mono-tri × tetra, from eight to 16 (average 13·5) plants were grown, and from one to 12 (average 3·9) from each cross were examined cytologically. The frequency of mono-tetras was thus in reasonable accordance with expectation on the basis of about three-fourths of the gametes of a monosomic plant being nullisomic and about one-third of the gametes of a trisomic carrying an extra chromosome.

The number of nulli-tetras recovered from selfed mono-tetra depends primarily upon the degree to which the extra chromosome compensates in the pollen for the missing one. Approximately three-fourths of the gametes are expected to be deficient for one chromosome

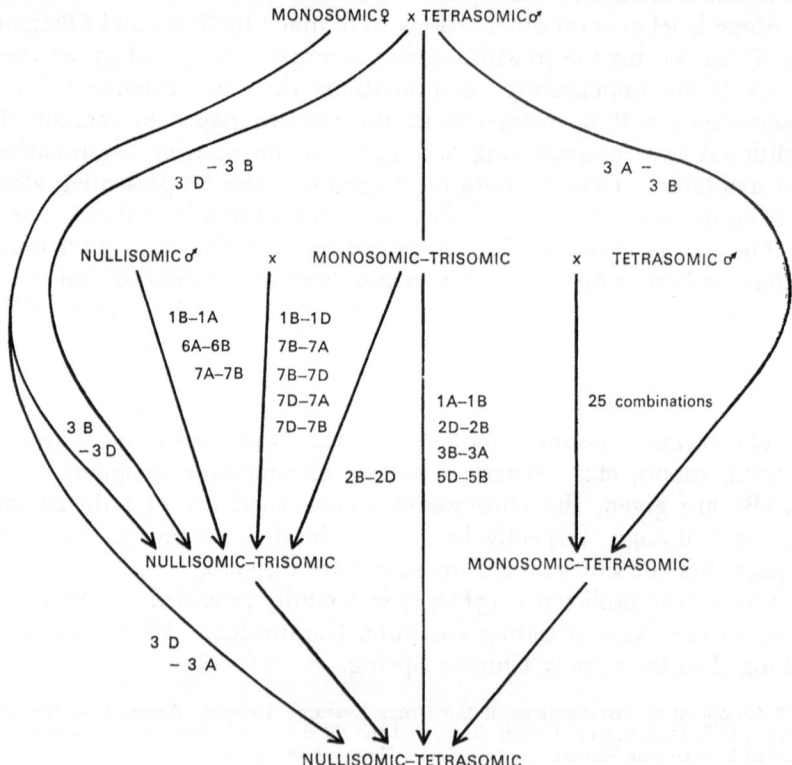

FIG. 1. Manner of origin of nullisomic-tetrasomics involving homoeologous chromosomes.

and duplicated for the homoeologue, assuming regular 2:2 distribution of the members of the tetrasome at meiosis, an assumption not realised with certain tetrasomes, notably those of group 3 (Sears, 1954). With no selection on the male side, this would result in about 9/16 nulli-tetra offspring. Because homoeologous chromosomes were involved, selection in favour of nulli-di pollen (n−1+1) could be expected in some cases. Since the offspring analysed cytologically (table 1) were usually not a random sample, no general conclusion can be drawn as

TABLE 1

Chromosome constitution of offspring of monosomic-tetrasomics

Chromosomes involved		No. grown	No. analysed	No. nulli-tetra
Mono	Tetra			
1A	1D	19	1	1
1D	1A	18	2	1
1D	1B	12	5	1
2A	2B	20	8	6
2A	2D	20	2	1
2B	2A	33	10	4
2D	2A	20	2	1
3A	3B	19	2	1
3A	3D	20	1	1
4A	4B	14	2	1
4A	4D	40	4	3
4B	4A	35	9	5
4B	4D	19	1	1
4D	4A	20	2	1
4D	4B	18	3	3
5A	5B	40	11	8
5A	5D	40	10	2
5B	5A	37	6	1
5B	5D	40	3	2
5D	5A	20	6	1
6A	6D	12	4	1
6B	6A	20	2	1
6B	6D	39	3	2
6D	6A	20	2	2
6D	6B	20	7	7
7A	7D	23	1	1
7B	7A	20	10	7

to the amount and kind of pollen selection that occurred. In nearly every case, the first few plants to reach the sampling stage were the ones analysed. In some cases, particularly those involving nulli-6D, this undoubtedly involved a selection in favour of nulli-tetras; whereas in other cases the nulli-tetras tended to be late and thus were poorly represented in the sample taken. In a few families—for example, those involving mono-4B and -6B—the nullisomics could be recognised by their awnedness, and the number of plants examined was thereby reduced.

The few data obtained from selfing of monosomic-trisomics conform to the conclusion previously reached (Sears, 1944) for mono-2D tri-2B and mono-3D tri-3A that nulli-di pollen may compete fairly successfully with normal pollen—that is, that an extra chromosome can compensate to a large extent for a missing homoeologue in the pollen as well as in the plant itself. The only significant addition to the 1944 data is for mono-2B tri-2D, which yielded three nulli-tetra, eight nulli-tri, and one normal in 34 plants analysed. As with 2D-2B, this is in reasonable accordance with prediction based on competition on even terms between normal and nulli-di pollen.

TABLE 2

Chromosome constitution of offspring of nullisomic-trisomics

Chromosomes involved		No. grown	No. analysed	No. nulli-tetra	No. nulli
Nulli	Tri				
1B	1A	20	20	9	1
1B	1D	20	2	1	0
2B	2D	40	27	5	3
2D	2B	104	62	29	0
3B	3D	22	8	1	2
3D	3B	19	1	1	0
6A	6B	13	6	2	0
7A	7B	51	29	5	4
7B	7A	22	13	1	2
7B	7D	20	10	1	3
7D	7A	9	9	2	0
7D	7B	20	4	1	0

Another measure of the degree of compensation of one chromosome for another in the pollen is the frequency of nulli-tetra and nullisomic offspring obtained from nulli-tri plants (table 2). If compensation is complete, nulli-tri plants should behave like the corresponding monosomics, except for differences based on the divergent meiotic behaviour of trisomes and monosomes. Whereas monosomics produce only about 25 per cent. of 21-chromosome gametes, a nulli-tri individual should give rise to about 40 per cent. of 21-chromosome (19+2) gametes. This means a ratio of 40:60 of 21- to 20-chromosome pollen instead of 25:75 and should result in the functioning of only about half as much 20-chromosome pollen. Also, only 60 per cent. of the functioning 20-chromosome pollen of the nulli-tri will give rise to nullisomic plants, compared with 75 per cent. from monosomics. Therefore, assuming full compensation, only about 0·4 per cent. to 3·0 per cent. nullisomics would be expected instead of the 0·9 per cent. to 7·6 per cent. reported (Sears, 1954) for monosomics. Nulli-tetras, on the other hand, would be expected to be more frequent than are disomics from monosomics—about 40 per cent. instead of 25 per cent.

For the combination 2D-2B (table 2), it is clear that nulli-tri pollen functions almost to the exclusion of nulli pollen, and the same may be true for 1B-1A. In 2B-2D, 3B-3D, 7A-7B, 7B-7A and 7B-7D, the rather meagre data suggest that nulli pollen competes fairly successfully with nulli-di, and this indicates less than complete compensation in the male gametophyte.

(ii) Characteristics

Most of the 42 nulli-tetra combinations involving chromosomes of the same homoeologous group are far superior to the corresponding nullisomics, and all are superior in some way.

TABLE 3

Number of seeds obtained on single plants having compensating nullisome-tetrasomes

Nullisome involved	Tetrasome involved	Season grown	No. seeds	Tetrasome involved	Season grown	No. seeds
1A	1B	1953S*	183	1D	1953S	425
1B	1A	1951S	118	1D	1951F	127
1D	1A	1957F	40	1B·	1954S	97
2A	2B	1951F	109	2D	1952F	129
2B	2A	1959F	17	2D	1951S	8
2D	2A	1952S	211	2B	1951S	129
3A	3B	1952S	118	3D	1951F	148
3B	3A	1953S	190	3D	1951S	203
3D	3A	1951S	117	3B	1951S	136
4A	4B	1955F	0	4D	1954F	0
4B	4A	1959F	35	4D	1954S	90
4D	4A	1952S	124	4B	1959F	7
5A	5B	1951F	58	5D	1951F	78
5B	5A	1959F	25	5D	1951F	63
5D	5A	1955F	53	5B	1963F	139
6A	6B	1951S	128	6D	1952F	162
6B	6A	1959F	47	6D	1959F	17
6D	6A	1951F	102	6B	1951F	107
7A	7B	1951S	62	7D	1953S	187
7B	7A	1951F	91	7D	1952S	229
7D	7A	1951S	91	7B	1951F	144

* S = spring; F = fall.

One of the major criteria used in establishing the superiority of the nulli-tetra to the nullisomic was fertility as measured by seed set from selfing. Since the plants scored for fertility were not all grown at the same time (table 3), the values obtained involve more than the usual amount of error due to environmental differences. Also, in some cases the most fertile plant was chosen from among several nulli-tetras, while in 15 cases only a single nulli-tetra plant was grown. Nevertheless, the plants were all raised in a greenhouse under such similar conditions that substantially the same seed sets were obtained on normal plants from season to season and year to year. Large differences in the recorded seed sets almost certainly reflect real differences in fertility.

Many of the nulli-tetras are of nearly normal fertility. It is an unusually fertile plant of normal constitution that has as many as 500 seeds, and 300 would be closer to the average value for plants handled as the nulli-tetras were (*i.e.*, grown in 6-inch pots until meiosis could be studied and then transferred to 8-inch pots). Thus any plant that bears as many as 200 seeds is approaching normal fertility, and a set of 100 seeds represents a fair degree of fertility.

In homoeologous group 3, all six nulli-tetra combinations had more than 100 seeds. In three other groups (1, 2 and 6), all exceeded 100 except the two combinations involving a particular nullisome (1D, 2B and 6B). Group 4 appears to have the poorest fertility, with the nulli-4A combinations completely sterile and 4D-4B nearly sterile. The group-5 combinations are consistently low, except for 5D-5B.

In comparison with the respective nullisomics, only the combinations 4A-4B, 4A-4D, and possibly 7B-7A and 7D-7A showed no increase in fertility. In about half the combinations, the increase over the nullisomic is particularly obvious because the nullisomics concerned set no selfed seeds (Sears, 1954). This is true of all three nullisomics of group 2, which are female-sterile; of groups 4 and 5, which are male-sterile; and of nullisomic 6B, which is male-sterile. Several other nullisomics, 1A, 1D, 3D, 6A and 6D, are only marginally fertile, and nullisomics 3A and 3B are poorly fertile. Nulli-7A has not been observed to set as many as 60 seeds, but nullisomics 7B and 7D have set as high as 112 and 126, respectively.

To supplement the data in table 3 and the information obtainable from the photographs of nullisomics and nulli-tetras in plates 1, 2 and 3, the following additional observations concerning the various groups are presented.

Group 1. The compensation is excellent, the spike peculiarities of the nullisomics (long internodes, stiff glumes) being fully suppressed in the nulli-tetras. In the one combination of low fertility (1D-1B), the plant is of nearly normal size.

Group 2. Here the compensation of 2A and 2D for each other, and that of 2B for 2A and 2D, are very good, but 2A and 2D compensate only partially for 2B. Not only is fertility low in these two combinations, but certain of the floral abnormalities of nulli-2B remain. Tetra-2A largely but not completely corrects the tendency of nulli-2B toward reduplication of floral parts. Tetra-2D restores nearly the normal number of spikelets to nulli-2B, but it does not restore normal internode length. This results in a long, lax spike. Both tetrasomes largely

PLATE I (*opposite*). Nullisomic-tetrasomics within homoeologous groups 1, 2 and 3. Each series of three spikes includes the simple nullisomic (left) and the two combinations of this nullisomic with homoeologous tetrasomes. Combination nulli-A tetra-B is to the left of A-D, B-A to the left of B-D, and D-A to the left of D-B. ×4/5.

1A

1B

1D

2A

2B

2D

3A

3B

3D

correct the coarseness and reduced tillering of nulli-2B. Full compensation in all combinations is exhibited for papery glumes and reduced awns.

Group 3. The degree of compensation in this group is probably best of all, particularly when plant characters are taken into account. Whereas all three nullisomics are narrow-leaved dwarfs, all six nulli-tetras are of essentially normal size and vigour. Even the tendency of nulli-3B toward asynapsis is less pronounced in the nulli-tetras.

Group 4. Nulli-4A and -4D are very similar in plant and spike characters, while -4B is narrower-leaved, bushier, and shorter in plant and spike. Tetra-4D does not compensate for the infertility of nulli-4A, however, although the plant and spike become decidedly more nearly normal. In the reverse direction compensation of tetra-4A for nulli-4D is the best in the group. Tetra-4B compensates poorly for both nulli-4A and -4D, but in part this is attributable to a necrotic condition that affects the leaves of tetra-4B and is independent of the dosage of 4A and 4D. Trisomic-4B does not show this necrosis, and a nulli-4A tri-4B plant obtained was more vigorous than the nulli-tetra, but it also set no seed. It appears that chromosome 4A carries a gene or genes essential for male fertility not present on the other two chromosomes, whereas 4B contributes more to length of spike than do the other two chromosomes. Since the hooded gene of 4B is not duplicated on 4A or 4D, neither of the latter two tetrasomes is able to suppress the awnedness of nulli-4B.

Group 5. Results with this group are confounded by the large effect of the gene Q carried by 5A. Although the spike characters of nulli-5A are changed very little by tetra-5B or -5D, tetra-5A spikes are considerably less compact in combination with the nullisomes, especially 5D. The aberrant plant characters, particularly the narrow leaves and slender culms, of all three nullisomes are substantially corrected by addition of the tetrasomes. Under certain conditions, both nulli-5D tetra-5A and nulli-5D tetra-5B show a pronounced tendency toward asynapsis.

Group 6. With respect to fertility it is clear that chromosome 6B carries a gene or genes not present on 6A or 6D, since the latter two tetrasomes do not restore full fertility to nulli-6B. Plant characters are reasonably normal in all the nulli-tetras. Tetrasome 6D is unable to restore normal spike length to nulli-6A and -6B, showing that 6A and 6B carry genes essential to normal spike length that are not present on 6D. As is already well known, the awn inhibitor B_2 on 6B has no duplicates on 6A and 6D; hence both nulli-tetras involving nulli-6B are awned.

Group 7. This is the group in which least compensation could be expected, because the nullisomics, particularly 7B and 7D, are themselves so nearly normal. It is obvious, however, from the fertility data

PLATE II (*opposite*). Nullisomic-tetrasomics within homoeologous groups 4, 5 and 6, together with the respective simple nullisomics and a normal spike (N). ×4/5.

and the photographs of spikes (plate 3, row 1) that both tetra-7B and -7D compensate strongly for nulli-7A, and that tetra-7D gives essentially full compensation for nulli-7B. Tetra-7A can also be said to compensate for nulli-7D, although the spike of nulli-7D photographed is neither quite as large nor as fertile as some that have been obtained. The spike shown of nulli-7D tetra-7B bears no more seeds than can be obtained on individual spikes of nulli-7D, but the 144 seeds set on the nulli-7D tetra-7B plant in table 3 are more than have been recorded for nulli-7D. The remaining combination, nulli-7B tetra-7A, is no less fertile than nulli-7B; it makes a taller plant; and its spikes are almost completely normal, whereas those of nulli-7B tend to be sterile in the upper portion. Therefore it may be concluded that all the nullisomic-tetrasomics in this group exhibit compensation, although necessarily of a lower order than that in some of the other groups.

TABLE 4

Chromosome constitution of offspring of nullisomic-tetrasomics

Chromosomes involved		No. grown	No. nulli-tetra	No. nulli-tri
Nulli	Tetra			
1B	1A	3	3	0
1D	1A	5	4	1
2B	2D	1	1	0
2D	2B	7	7	0
3A	3B	2	2	0
3B	3D	5	5	0
3D	3A	11	7	4
3D	3B	2	2	0
4B	4A	2	2	0
4D	4B	3	3	0
5B	5A	3	2	1
5B	5D	4	4	0
5D	5B	6	3	3
6A	6B	3	3	0
6A	6D	1	1	0
6B	6A	3	3	0
6B	6D	3	3	0
7A	7B	3	1	2
7B	7A	2	2	0
7D	7A	3	2	1
Totals		72	60	12

(iii) *Breeding Behaviour*

From cytological analysis of offspring of about half the compensating combinations (table 4), scattered through the seven homoeologous groups, it is clear that this type of nullisomic-tetrasomic forms a reasonably stable line. Of the 72 offspring 60 were nulli-tetra, and the remaining 12 were nulli-tri. This kind of stability was to be expected

TABLE 5

Nullisomic-tetrasomic (or otherwise deficient-duplicated) combinations obtained involving chromosomes of different homoeologous groups

Groups involved	Genomes involved							
	A-A	A-B	A-D	B-B	B-A	B-D	D-A	D-B
1-2						X		
1-3					X			
1-4					X			
1-5						X		
1-6				X	X			
1-7					X		X	
2-1							X	
2-3			X					
2-4			X					
2-5		X						
2-6	X	X						
2-7				X	X			
3-1			X					
3-2			X					
3-4	X							X
3-5			X					
3-6		X						
3-7			X					
4-1		X						
4-2			X	X				
4-3	X							
4-5				X	X	X		X
4-6								X
4-7			X					
5-1	X	X						
5-2			X					
5-3				X	X			
5-4		X		X			X	
5-6		X						
5-7	X		X					
6-1							X	
6-2	X							
6-3						X		
6-4			X					
6-5			X					X
6-7	X				X			
7-1		X	X					
7-2	X	X						
7-3			X					
7-4			X					
7-5			X					
7-6	X	X						
	9	10	16	6	8	4	4	4

because of the competitive advantage 21-chromosome (nulli-di) pollen has over 20-chromosome pollen when the duplicated chromosome tends to compensate for the missing one. The occurrence of occasional nulli-tri plants must be very largely a consequence of irregular meiotic behaviour of the tetrasome on the female side. Failure of one or more members of the tetrasome to pair, which would usually be followed by univalent loss, must be a common type of irregularity, for simple 3:1 distributions from quadrivalents would lead to as many pentasomes as trisomes, and no pentasomes were observed in the sample of 72.

3. NON-COMPENSATING COMBINATIONS

The ability of each tetrasome to compensate to some extent for each of the two nullisomes of the same homoeologous group shows that the three chromosomes of each group have most of their genes in common. The possibility remains, however, that considerable relationship exists between certain chromosomes of different homoeologous groups.

To check every tetrasome for its ability to compensate for every non-homoeologous nullisome would require 378 tests, an inordinately large number. If one tetrasome from each homoeologous group is tested, however, with one nullisome from each other group, only 42 tests are necessary, and this perhaps constitutes a reasonably representative sample of the combinations possible.

Since some of the weaker nullisomics might become inviable if burdened additionally with a non-compensating tetrasome, the chromosome chosen to make nullisomic was in most cases the one whose nullisomic was the most vigorous in its homoeologous group. An exception was group 7, where nullisomics 7A and 7D are so nearly normal that a slight increase in vigour or fertility resulting from addition of a partially compensating tetrasome might not be detectable. In group 2, 2A was chosen because nulli-2A is less coarse than either nulli-2B or -2D; and since most of the tetrasomics of other groups tend to be coarse, their combinations with nulli-2B or -2D might be so excessively coarse as to conceal points of improvement.

Although nulli-3B is no less vigorous than -3A, it was not used because chromosome 3B carries a gene essential to normal synapsis. Nulli-4A is more vigorous than -4B, but the latter was used with three different tetrasomes of group 5 because its morphology suggested possible affinities with that group. Nulli-6D is somewhat more vigorous than -6A or -6B, but the latter two were also used.

PLATE III (*opposite*). Nullisomic-tetrasomics of homoeologous group 7, the respective group-7 nullisomics, a normal spike (N), and various non-compensating, between-group combinations (centre and bottom rows). Spikes in the centre row are from nullisomic-tetrasomics as follows: (1) 2A-4D, (3) 4B-5A, (4) 5D-4A, (5) 6D-1A, (6) 7A-1B, (7) 7A-4D, (8) 7A-6B, and (2) nullisomic-trisomic 3A-4A. In the bottom row are monosomic-tetrasomics (1) 1B-6A, (2) 1D-7A, (3) 2A-5B, (4) 2A-6A, (5) 2B-7A, (6) 3A-2D, (7) 4A-7D, (8) 5A-2D, (9) 6B-7A. ×4/5.

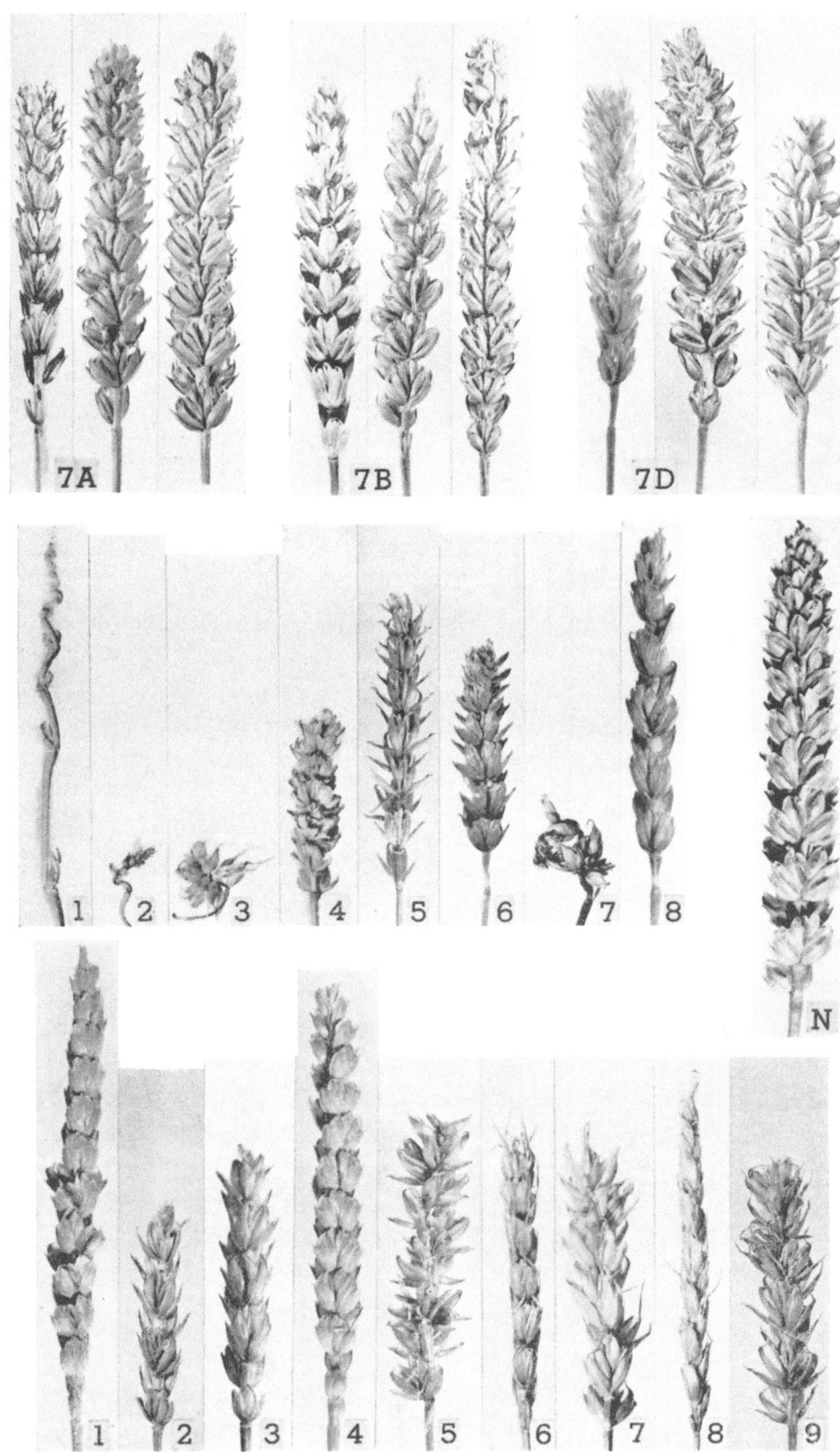

to face p. 38

Four combinations were tested because Okamoto and Sears (1962) found translocations from haploids that involved the chromosomes concerned. The translocations were thought to have arisen through pairing and crossing over, which would mean structural, and presumably genetic, relationship. These combinations were 2B-4B, 3A-4A, 5A-7A and 5D-6A.

In all, 61 combinations were tested (table 5).

The method used for producing the between-group nulli-tetras was the same as for the majority of the within-group combinations; namely, (1) cross monosomic by tetrasomic, (2) backcross by tetrasomic, and (3) self the resulting mono-tetra.

As a rule little difficulty was experienced in obtaining the mono-tetra, as expected since there should be no selection of gametes in mono-tri × tetra. With tetrasomics of group 3, however, there was poor recovery of mono-tetras, which is presumably related to the pronounced tendency of tetrasomes of this group to revert to trisomes (Sears, 1954). In the nine combinations attempted involving group-3 tetrasomes, an average of 17·4 plants from mono-tri × tetra had to be examined to find one mono-tetra; whereas in 57 combinations involving tetrasomes of the other groups, only 4·4 plants were examined per mono-tetra.

In many cases it was clear from the characteristics of the mono-tetra that the tetrasome was not compensating for the missing chromosome but was making the plant more abnormal. This was particularly obvious in the effect on fertility, many mono-tetras setting few or no seeds (table 6), whereas the respective monosomics are fully fertile or nearly so. Spikes of several infertile mono-tetras are shown in plate III, bottom row, figs. 3, 4, 6, 7, 8. The remaining spikes in this row have a low degree of fertility.

Where seeds were produced by the mono-tetra, progeny were grown in an effort to obtain the nulli-tetra. In most cases this effort failed, in spite of some populations being sizeable (table 7). Although relatively few plants were examined cytologically, it is unlikely that nulli-tetras were overlooked. All seedlings with characteristics suggesting nullisomics were saved for cytological analysis. In several combinations the apparent nullisomic died before reaching the sampling stage, strongly suggesting that it was a semi-lethal nulli-tetra. In a few cases the nulli-like plants survived but produced spikes so abnormal that meiotic stages could not be obtained (plate III, figs. 1, 3, 7 in centre row). These and the pre-heading lethals are recorded as probable nulli-tetras in table 6. Of the remaining nulli-tetra spikes figured, all were sterile except no. 8, which set a single seed.

In two combinations (2A-6B and 5A-6B) no nulli-tetra was obtained, but a monotelosomic-tetrasomic did appear. In 3A-1D a monotelosomic-trisomic was obtained. In each of these cases no superiority was shown to the corresponding nullisomic. This indicated a lack of compensation by the tetrasomes or trisome, since without the extra

TABLE 6

Chromosomes involved in the non-compensating combinations,
and kind of combination obtained

Groups involved	Combinations obtained and chromosomes concerned				
	Nulli-tetra	Probable nulli-tetra	Nulli-tri	Mono-tetra	
				With no seeds	With few seeds
1-2				1B-2D	
1-3				1B-3A	
1-4				1B-4A	
1-5					1B-5D
1-6				1B-6A, 1B-6B	
1-7				1B-7A, 1D-7A	
2-1				2D-1A	
2-3				2A-3D	
2-4		2A-4D			
2-5				2A-5B	
2-6				2A-6A	2A-6B
2-7				2B-7B	2B-7A
3-1			3A-1D		
3-2				3A-2D	
3-4			3A-4A	3D-4B	
3-5		3A-5D			
3-6			3A-6B		
3-7				3A-7D	
4-1		4A-1B			
4-2					4A-2D, 4B-2B
4-3			4A-3A		
4-5	4B-5D	4B-5A, 4B-5B		4D-5B	
4-6					4D-6B
4-7				4A-7D	
5-1			5A-1B		5A-1A
5-2				5A-2D	
5-3					5B-3A, 5B-3B
5-4	5D-4A			5A-4B, 5B-4B	
5-6	5A-6B				
5-7				5A-7D	5A-7A
6-1				6D-1A	
6-2		6A-2A			
6-3					6B-3D
6-4					6A-4D
6-5					6A-5D, 6D-5B
6-7					6A-7A, 6B-7A
7-1	7A-1B, 7A-1D				
7-2			7A-2A		7A-2B
7-3					7A-3D
7-4		7A-4D			
7-5					7A-5D
7-6	7A-6B				7A-6A

chromosomes the monotelosomic should have been superior to the nullisomic.

Where numerous offspring were grown from the monosomic-tetrasomic but no nulli-tetras were obtained, it was assumed either

TABLE 7

Chromosome constitution of offspring of monosomic-tetrasomics

Parents		No. grown	Offspring							
			Constitution of plants analysed							
Mono	Tetra		Nulli-tetra (or tri)	Monotelo-(or -iso-) tetra (or tri)	Mono-tetra	Tetra	Mono-tri	Tri	Mono	Nulli
1B	5D	16			5	2		2		
1B	6A	14			4	3				
1B	6B	67			8	2	4	1		
1B	7A	2			2					
1D	7A	2			1					
2A	4D	12			3		1			
2A	6B	131		1	13	5	1	1	1	
3A	1D	7		1					1	
3A	4A	79	1		7		1			
3A	5D	48	1		5		1			
3A	6A	40			2				2	
3A	6B	29	1							
3A	7D	1			1					
3D	4B	11			1	1				
4A	1B	102			6	5	1			
4A	2D	6			1	5				
4B	5A	130			12	8	5	1		1
4B	5B	319			16	6	1			
4B	5D	74	1	1	1					
4D	6B	15			2					
5A	1A	4			3					
5A	1B	30			1		1			
5A	6B	70		1						
5D	4A	34	1		1					
6A	2A	158			14	1	1			
6A	4D	131			17	7	3	1		
6B	3D	32								
6D	1A	77	3		2					
6D	5B	149			17	8	4	2		
7A	1B	35	2	1	3					
7A	1D	33	2							
7A	2A	117	2		16	8		2		
7A	4D	39			1		1			
7A	6B	25	1		3					

that male transmission of the nulli-di gametes was very low or that the nulli-tetra individuals produced were lethal at a very early stage. In either case it seemed very unlikely that the extra chromosome was

D

compensating for the missing one. As a further test, however, a comparison was made of the mono-tetra with the corresponding tetrasomic. If the mono-tetra was inferior, it seemed certain that the nulli-tetra would be still poorer, and therefore that no detectable compensation was involved. As a sensitive and convenient measure of vigour, the number of seeds set per plant was used. In no case was the mono-tetra more fertile than the corresponding tetrasomic.

In addition to the bizarre interactions of nullisomes and tetrasomes figured in plate III, tetrasome 7A affected mono-2B in an unexpected way. Both these two chromosomes are essential to normal floral development (Sears, 1954), with loss of 2B leading to a tendency for extra flowers or bracts to be inserted between glumes, and loss of 7A resulting in pistillody. In mono-2B tetra-7A, which has extra, not reduced, dosage of 7A, extreme pistillody, leading to almost complete sterility, was observed.

4. DISCUSSION

The results of the nullisomic-tetrasomic tests were unequivocal in showing some degree of compensation in each of the 42 combinations of chromosomes from the same homoeologous group and no compensation in any of the 61 tests of chromosomes from different groups. This would suggest that the main process by which the chromosomes of the A, B and D genomes have become structurally differentiated from each other is inversion. Inversions simply rearrange the chromatin within the individual chromosome and do not change the gene content. By this process homologous chromosomes in different genomes can become homoeologous.

Although combining tetrasomes with nullisomes revealed no relationships outside the homoeologous groups, this is a rather crude test, only capable of revealing substantial amounts of homology. If a tetrasome possesses relatively few of the same genes as the nullisome with which it is being combined, it restores the dosage of these genes but at the same time supplies an overdose of the rest of its own genes, with a net effect which may be detrimental. It is thus possible that considerable homology does exist between chromosomes of different homoeologous groups. Such homology would be most likely to arise through the process of reciprocal translocation.

In groups 1 and 3, the excellent within-group compensation indicates that the homoeologues have essentially the same genetic content. Therefore it is very unlikely that any substantial amount of homology exists between any chromosomes of these groups and non-homoeologous chromosomes. Although groups 5 and 7 show a somewhat lower level of compensation within themselves, no one chromosome is particularly anomalous, and there seems to be only little likelihood of a significant degree of homology with any non-homoeologous chromosomes. In each of groups 2, 4 and 6, however, there is one chromosome, 2B, 4A and 6B, for which the others compensate poorly.

This suggests that each of these chromosomes may have a segment, presumably derived from some non-homoeologue, that is not possessed by its two homoeologues.

One way of explaining this situation is to assume that 2B, 4A and 6B were involved in a double translocation, such that, for example, 2B had a segment replaced by one from 4A, 4A had a segment from 6B, and 6B had the segment from 2B. In this case nulli-2B would not be completely compensated for by tetra-2A or -2D, because neither of these would supply the missing 4A-segment. Neither would 4A be able to supply this segment; but 4B and 4D would have the segment and might show compensation for 2B. Similarly, tetra-6A and -6B would tend to compensate for nulli-4A, and tetra-2A and -2D for nulli-6B. The reciprocal combinations might also show compensation.

That the situation is not so simple as this is indicated by the fact that in each group the anomalous chromosome compensates very well as a tetrasome for each of the other two nullisomes. This would not be the case if it had lost an important segment through reciprocal translocation with a non-homoeologue. It is of interest to note, however, that one of the translocations obtained by Okamoto and Sears (1962) from haploids involved chromosomes 2B and 4B. If, as the authors assumed, these translocations arose through pairing and crossing-over, then 2B has a segment in common with 4B—a segment which 2B could have acquired from 4A.

Another translocation of Okamoto and Sears involved chromosome 4A, but the second chromosome was 3A, which belongs to a homoeologous group in which compensation is so nearly complete that substantial relationship of any of its three members to a member of another group seems unlikely. The other two non-homoeologous translocations from haploids involved 5A with 7A and 5D with 6A. These chromosomes belong to less well-defined homoeologous groups than does 3A, and thus relationship between them is not so improbable. Chromosomes 7A and 5D failed to compensate for 5A and 6A, respectively, but the reverse combinations were not tested. It is of course possible that these chromosomes have no common segment—that the non-homoeologous translocations of Okamoto and Sears originated through some process other than pairing and crossing over.

Although chromosomes 2B, 4A and 6B were included in as many or more compensation tests than the average for the other 18 chromosomes, additional tests of these three anomalous chromosomes would be desirable. However, it may be more productive to use a cytological method for checking possible relationships involving not only these chromosomes but also such chromosomes as the ones just mentioned that underwent translocation in haploids. A promising cytological method was indicated by the discovery by Okamoto (1957) and Riley (1958) of the suppressing effect which chromosome 5B exerts on homoeologous pairing. With chromosome 5B missing (or its effect suppressed), pairing may presumably be obtained between chromosomes with only a small segment in common. Riley and Kempanna (1963) did not detect any pairing of non-homoeologues in nulli-5B material, but in their experiment pairing of homoeologues had to compete with pairing of homologues. To test for the possibility of pairing between two particular chromosomes, the ideal situation will be to have both chromosomes monosomic. Then neither will have a homologue with which to pair, and their homoeologues, being present as disomes rather than monosomes, will offer a minimum of pairing competition.

It is perhaps worth noting that two of the three anomalous chromosomes, 4A and 6B, are considerably longer than their homoeologues. This suggests that in these two cases the homoeologues have simply become deficient for a segment which 4A and 6B still retain. This would account for the fact that the tetrasomes of these two chromosomes are able to compensate well for the nullisomes of their homoeologues, whereas compensation is poor in the reverse direction. It would also explain that 6B is a nucleolar-organising chromosome while 6A and 6D are not. The behaviour of chromosome 2B cannot be explained in this way without some additional assumptions, for 2B is definitely shorter than 2A and little, if any, longer than 2D (Morrison, 1953; Sears, 1954).

The extent to which homology appears to be confined to the homoeologous groups suggests that differentiation of the chromosomes of the three genomes of wheat has involved very few translocations. Yet at least five different translocations with respect to the variety used here, Chinese Spring, have been identified in crosses of a relatively small number of varieties with monosomics of Chinese Spring (Sears, unpublished). Is it possible that translocations have been an important factor in the differentiation of the wheat chromosomes, but that the variety Chinese Spring happens to have a chromosome architecture which is more primitive than that of most other varieties? The answer is not necessarily in the affirmative, as a few translocations could in fact help to account for some of the anomalies reported here in the compensation tests within homoeologous groups. However, the data of Riley and Chapman (1960) show that at least the D genome of Chinese Spring does have a primitive arrangement. In a hybrid of this variety with *Aegilops squarrosa*, the source of the D genome, they found usually seven bivalents and never any multivalents.

5. SUMMARY

Combinations of tetrasomes with nullisomes showed that each chromosome of hexaploid wheat has a close relative in each of the other two genomes, making seven homoeologous groups of three. Within these groups all 42 possible combinations of tetrasomes with nullisomes have been made, and in every case the tetrasome compensates to some extent for the deleterious effect of the nullisome. Compensation is excellent in groups 1 and 3, whereas the tetrasomes homoeologous to nullisomes 2B, 4A and 6B compensate rather poorly for them. It is suggested that the anomalous behaviour of these three chromosomes may be the result of translocation involving them; or that, in the case of 4A and 6B, a deficiency may have occurred involving their homoeologues.

At least one nullisome or monosome from each homoeologous group was combined with at least one tetrasome or trisome from each other group, for a total of 61 combinations, none of which gave evidence of compensation. Only 19 could be obtained as nullisomic-tetrasomic (or -trisomic), the rest being evaluated as monosomic-tetrasomic. Although the tests established that genetic homology is very largely confined to the homoeologous groups, they did not exclude the possibility that small amounts of homology, undetectable by the relatively insensitive nullisomic-tetrasomic test, may exist between certain non-homoeologues.

6. REFERENCES

MORRISON, J. W. 1953. Chromosome behaviour in wheat monosomics. *Heredity*, 7, 203-217.

OKAMOTO, M. 1957. Asynaptic effect of chromosome V. *Wheat Information Service*, 5, 6.

OKAMOTO, M. 1962. Identification of the chromosomes of common wheat belonging to the A and B genomes. *Can. J. Genet. Cytol.*, 4, 31-37.

OKAMOTO, M., AND SEARS, E. R. 1962. Chromosomes involved in translocations obtained from haploids of common wheat. *Can. J. Genet. Cytol.*, 4, 24-30.

RILEY, R. 1958. Chromosome pairing and haploids in wheat. *Proc. 10th Int. Congr. Genet.*, 2, 234-235.

RILEY, R., AND CHAPMAN, V. 1960. The D genome of hexaploid wheat. *Wheat Information Service*, 11, 18-19.

RILEY, R., AND KEMPANNA, C. 1963. The homoeologous nature of the non-homologous meiotic pairing in *Triticum aestivum* deficient for chromosome V (5B). *Heredity*, 18, 287-306.

SEARS, E. R. 1944. Cytogenetic studies with polyploid species of wheat. II. Additional chromosomal aberrations in *Triticum vulgare. Genetics*, 29, 232-246.

SEARS, E. R. 1952. Homoeologous chromosomes in *Triticum aestivum. Genetics*, 37, 624.

SEARS, E. R. 1954. The aneuploids of common wheat. *Res. Bull., Mo. agr. Exp. Stn.*, 572, 59 pp.

SEARS, E. R. 1958. The aneuploids of common wheat. *Proc. 1st Int. Wheat Genet. Symp.*, 221-228.

SEARS, E. R., AND OKAMOTO, M. 1956. Genetic and structural relationships of non-homologous chromosomes in wheat. *Proc. Int. Genet. Symp. (Cytologia Suppl. Vol.)*, 332-335.

ESTIMATES OF THE HOMOEOLOGY OF WHEAT CHROMOSOMES BY MEASUREMENTS OF DIFFERENTIAL AFFINITY AT MEIOSIS

RALPH RILEY and VICTOR CHAPMAN
Plant Breeding Institute, Cambridge, England

1. INTRODUCTION

THE chromosome complement of the hexaploid wheat of commerce, *Triticum aestivum* ($2n = 6x = 42$), can be classified into seven homoeologous groups, each of three pairs, or into three genomes, each of seven pairs. The genomes represent the sets of chromosomes combined together—from three distinct diploid species—during the allopolyploid evolution of wheat. The 14-chromosome, diploid, ancestors of *T. aestivum* were *Triticum monococcum, Aegilops speltoides* and *Aegilops squarrosa,* or their close relatives—the contributors of the A, B and D genomes respectively (McFadden and Sears, 1944; Riley, Unrau and Chapman, 1958).

The three pairs of chromosomes in each homoeologous group perform similar genetic functions. Thus, although all the 21 different nullisomics of *T. aestivum* are phenotypically abnormal, plants that are simultaneously nullisomic for one and tetrasomic for another chromosome are of more or less normal phenotype, provided that the chromosomes in altered dosage are homoeologous. By contrast there is no compensation for the defects caused by nullisomy when the tetrasomic and nullisomic chromosomes are in different homoeologous groups (Sears, 1954, 1965).

Of the three homoeologous chromosomes in each group, one is in each genome. Consequently the correspondence in the genetic activities of homoeologues implies that they represent equivalent chromosomes derived from the different diploid parents of the hexaploid. Their equivalence may be presumed, therefore, to be due to their origin from the same chromosome of a remote common progenitor of all three ancestral diploid species.

Only bivalents are formed at meiosis in *T. aestivum* and there is disomic inheritance. Each chromosome therefore pairs only with its single fully homologous partner and there is normally no meiotic association between homoeologues. The absence of homoeologous pairing has been shown to be due to a genetic activity of a single chromosome—number 5B (Riley and Chapman, 1958; Riley, 1960; Riley and Kempanna, 1963). Homoeologous pairing occurs in plants deficient for this chromosome, although the deficiency of no other chromosome causes similar changes in the course of meiosis. The prevention of homoeologous pairing in the presence, and its occurrence in the absence, of chromosome 5B takes place in polyhaploid forms of *T. aestivum* as well as in normal hexaploid plants.

There is little pairing at meiosis in 28-chromosome hybrids between *T. aestivum* and a range of diploid species in the genus *Aegilops* because homoeologous synapsis is inhibited. This results from the continued activity of chromosome 5B, in the hybrids, and from the extension of its influence to the inhibition of the synapsis of *Aegilops* with wheat chromosomes (Riley, 1965; Riley and Law, 1965). By contrast there is a high level of meiotic pairing in hybrids between *T. aestivum* and the two diploid species, *Ae. speltoides* and *Ae. mutica*. Bivalents, trivalents and quadrivalents are common, although higher configurations are very rare, and there are some cells in which all 28 chromosomes have undergone synapsis (Riley, Kimber and Chapman, 1961; Riley, 1965). From behaviour of this type the hypothesis was formulated that the pairing is homoeologous and that consequently the normal inhibitory activity of chromosome 5B is suppressed in the presence of the genotype of either *Ae. speltoides* (Riley, Unrau and Chapman, 1958; Riley, Kimber and Chapman, 1961) or *Ae. mutica* (Riley, 1965).

The present work was initiated to test this hypothesis by the determination of the relationships of the chromosomes that paired in *T. aestivum* × *Ae. speltoides* hybrids. In the event, not only was the hypothesis shown to be correct, as has been indicated in a preliminary publication (Riley and Chapman, 1964a), but new detail was revealed about homoeologous relationships, and meiotic differential affinity was estimated quantitatively from cytological observations for the first time.

2. MATERIAL AND METHODS

The wheat plants used in the present work were all derivatives of *Triticum aestivum* L. emend. Thell. ssp. *vulgare* Mac Key variety Chinese Spring (2n = 6x = 42). All the lines employed were ditelocentric, having 20 pairs of normal chromosomes, with median or sub-median centromeres, and one pair of telocentric chromosomes. One arm of the chromosome marked by the telocentric condition was present in the normal double dose, while the other arm was completely deficient. The advantage of such material is that the marked chromosome is readily recognisable, by its distinctive morphology in both mitotic and meiotic metaphase preparations.

The chromosomes of *T. aestivum* are designated in a manner that indicates the homoeologous group and the genome to which they belong. The chromosomes of homoeologous group 1 are thus 1A, 1B and 1D, while those of the A genome are 1A, 2A . . . 7A. In the present paper a plant carrying a pair of telocentrics for one arm of a chromosome, for example 1A, and deficient for the other arm, will be designated *ditelocentric 1A*.

The other species used in the present study was *Aegilops speltoides* Tausch (2n = 14), in which the entire chromosome complement has median or sub-median centromeres. The taxonomic and evolutionary status of this species has most recently been considered by Zohary and Imber (1963).

Anthers from the plants studied were fixed in acetic-alcohol and stained by the Feulgen procedure, supplemented by the addition of propionic orcein. Analyses of chromosome pairing were made on permanent squashes of pollen mother cells at first metaphase of meiosis.

3. EXPERIMENTAL PROCEDURE

To determine the relationships of the chromosomes that pair in *T. aestivum* × *Ae. speltoides* hybrids it was necessary to introduce two structural markers simultaneously, since wheat chromosomes are indistinguishable from each other at meiosis. This was achieved by crossing together plants from lines in which different chromosomes were ditelocentric. The products of the cross had two pairs of chromosomes in which one member of the pair was a normal chromosome and the other was telocentric. These derivatives were pollinated by *Ae. speltoides* to produce 28-chromosome hybrids which were selected for the presence of two telocentric chromosomes. Such *T. aestivum* × *Ae. speltoides* hybrids carried the telocentrics of two different, but identified, chromosomes.

Hybrids were produced in which two chromosomes were simultaneously marked by a telocentric condition for all three combinations of chromosomes of homoeologous group 5. The marked chromosomes in these within-group combinations were thus 5A-5B, 5A-5D and 5B-5D. In order that the pairing behaviour of non-homoeologous chromosomes could be compared with that of homoeologues, an attempt was made to produce hybrids in which every chromosome of group 5 was telocentric in combination with every chromosome of groups 3 and 6. The eleven, out of sixteen possible, combinations of this type obtained are indicated below:—

	5A	5B	5D
3A	+	+	
3B	+	+	
3D	+	+	+
6A			
6B	+	+	
6D		+	+

It should be mentioned that the telocentric of chromosome 5B, used in this work, was for the arm responsible for the profound genetic effect on meiotic pairing. Consequently the wheat components of the genotypes of the hybrids were in no instance responsible for homoeologous conjugation.

4. CONFIGURATIONS INVOLVING MARKED CHROMOSOMES

(i) *Non-homoeologues*

In the *T. aestivum* × *Ae. speltoides* hybrids in which eleven different combinations of two non-homoeologous telocentrics were separately present the marked chromosomes were never observed in the same configuration. The telocentrics participated in bivalents, trivalents and quadrivalents, but always in different figures (plate I, fig. 1). Although very many meiotic cells with marked non-homoeologues were examined,

the failure to detect non-homoeologous pairing, like all negative observations, is of course not conclusive. It cannot be asserted categoric-
ally that the non-homoeologous chromosomes concerned never pair;
but it can be accepted without question that, if such pairing occurs,
it does so with extreme rarity.

(ii) Homoeologues

The meiotic behaviour of the marked chromosomes relative to
each other was quite different in the three types of *T. aestivum* × *Ae.*
speltoides hybrids with two homoeologous chromosomes telocentric.
The chromosomes marked by telocentric conditions in these hybrids
were either 5A and 5B, 5A and 5D, or 5B and 5D; and in every instance, in some first metaphase cells, these marked homoeologues
participated in the same configuration.

Some cells had both telocentrics unpaired, while in others both
paired with a complete chromosome in separate bivalents. In other
cases only one telocentric had paired and it was involved in either a
bivalent or a trivalent. These types of pairing were of course un-
instructive in terms of the relationships of the telocentrics concerned,
since they were also to be found in hybrids with non-homoeologues
marked.

The categories that were unique to hybrids with homoeologues
marked were those in which the two telocentrics were associated in the
same configuration. In the most commonly observed figure the two
telocentrics were paired together directly to form a rod-shaped bivalent
(plate I, figs. 2 and 3). This was the most direct and unequivocal distinction separating the meiotic behaviour of marked homoeologues from
that of non-homoeologues. However, there were also two other patterns
of pairing, found only when the telocentrics were homoeologous. The
two telocentrics were sometimes associated with a single complete
chromosome to form a triradial trivalent (plate I, fig. 4), or with two
complete chromosomes to form a chain-of-four quadrivalent with the
telocentrics at opposite ends. No matter which of the three combinations of two chromosomes of group 5 were telocentric, the same configurations were formed.

Since non-homoeologous telocentrics were never observed to pair,
the involvement of group 5 chromosomes in common configurations
provides clear confirmation of the hypothesis that meiotic pairing in
T. aestivum × *Ae. speltoides* hybrids is homoeologous. Furthermore,
because such pairing in haploids and hybrids is normally prevented
by the activity of the long arm of chromosome 5B (Riley, 1960; Riley
and Chapman, 1964*b*), which was present in all the hybrids, this
activity must be suppressed in the presence of the genotype of *Ae.*
speltoides.

The capacity of the telocentrics of 5A, 5B and 5D to pair, under the
present experimental conditions, provides cytological evidence that

reinforces the genetical criteria used by Sears (1954, 1965) in assigning these chromosomes to the same homoeologous group. In addition, it is apparent that the technique employed—using marked chromosomes and the suppression of the 5B effect—permits the unequivocal recognition of cytological homoeology. However, as will be seen later, while the overall pattern of genetical and cytological homoeology may be the same, there may be differences in detail.

(iii) *Arm relationships of homoeologues*

The nature of the figures in which the telocentrics of 5A, 5B and 5D participated together also provided evidence on the relationships of the arms of the chromosomes concerned. In every instance the telocentrics used in the present work paired together directly (table 1). That is, chiasma formation was possible between the arms that were telocentric, under conditions permitting homoeologous synapsis (plate I, figs. 2 and 3). Thus all three telocentrics represent corresponding arms of the homoeologues. This might indeed have been expected, since in all three cases the telocentric for the longer arm of a heterobrachial chromosome was employed. Nevertheless the present evidence is the first available on the arm relationships of homoeologues; and it shows that the arms with the following genes, or effects, are equivalent:

5A—Q (speltoid suppression), b_1 (recessive to dominant awn inhibitor).

5B—p_{5B} (inhibitor of homoeologous pairing).

5D—spring/winter growth habit.

5. DIFFERENTIAL AFFINITY OF HOMOEOLOGUES

There were striking contrasts in the frequencies with which the marked homoeologous chromosomes participated in different pairing configurations, depending upon the particular telocentrics present in the *T. aestivum* × *Ae. speltoides* hybrids (table 1). Thus, when 5A and 5B or 5A and 5D were telocentric, the direct pairing of the marked homoeologues—either in a bivalent or in a trivalent—was comparatively rare, occurring in only eight cells in one hundred in both combinations. By contrast, when 5B and 5D were marked the telocentrics

PLATE I (*opposite*). First metaphase of meiosis in pollen mother cells of *T. aestivum* × *A. speltoides* hybrids in which two chromosomes of *T. aestivum* are marked by a telocentric condition.

FIG. 1. There are two quadrivalents, one trivalent, seven bivalents and three univalents. The telocentrics are 3D, which is in a quadrivalent (arrowed), and 5D, which is in a bivalent (arrowed).

FIG. 2. One trivalent, 11 bivalents and three univalents; chromosomes 5B and 5D, which are telocentric, are paired together to form a bivalent (arrowed).

FIG. 3. One quadrivalent, one trivalent, six bivalents and nine univalents; chromosomes 5B and 5D are telocentric and paired together in a bivalent (arrowed).

FIG. 4. Two trivalents, five bivalents and 12 univalents; chromosomes 5B and 5D are telocentric and participate with a complete chromosome in a triradial trivalent (arrowed).

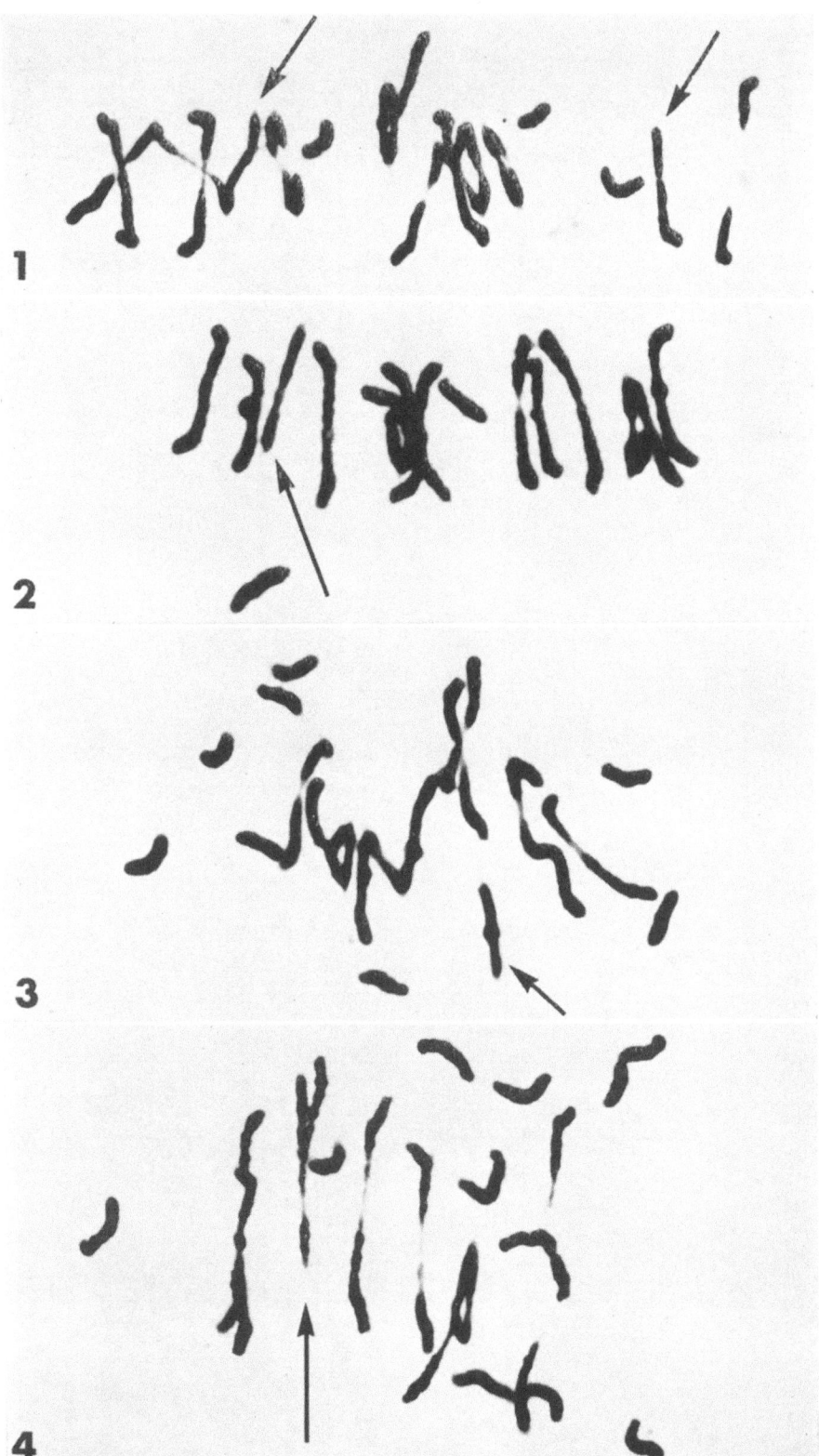

paired together directly in more than half the cells scored (table 1). In conditions permitting homoeologous pairing there was, therefore, pronounced meiotic differential affinity between the long arms of chromosomes 5A, 5B and 5D—5B and 5D displaying much closer affinity to each other than either did to 5A.

TABLE 1

Pairing behaviour of two telocentric chromosomes of homoeologous group 5 at MI of meiosis in T. aestivum × Ae. speltoides hybrids

Telocentric chrom.	No telo pairs	Telos together			One telo/complete				Two telo/complete			Total cells
				sum				sum			sum	
5A-5B	21	7	1	8	48	4	2	54	17	—	17	100
5A-5D	24	7	1	8	50	5	1	56	11	1	12	100
5B-5D	14	44	8	52	30	—	—	30	3	1	4	100
(1)	(2)	(3)	(4)		(5)	(6)	(7)	(8)	(9)	(10)	(11)	(12)

= telocentric chromosome

= complete chromosome

However, there were four chromosomes in each group capable of homoeologous pairing in the present hybrids, as is displayed by the occasional occurrence of quadrivalents (plate I, figs. 1 and 3) (table 1). Pairing may be presumed to be possible between the three chromosomes of group 5 derived from *T. aestivum* and between the corresponding chromosome of *Ae. speltoides*, which may be called 5S. In the largest configurations—quadrivalents—chromosomes 5A, 5B, 5D and 5S must all participate, and similar quadrivalents must be visualised involving the four chromosomes of every other group.

The relative affinities of each of the group 5 chromosomes derived from *T. aestivum* have already been discussed, but the recognition that four, not three, chromosomes were capable of homoeologous conjugation means that the relative affinities of each chromosome for three others can be separately estimated. It is thus possible to consider meiotic affinities in six two-by-two combinations of the group 5 chromosomes. These may be expressed as follows with the letters a, b, c, x, y and z representing the frequency with which two particular chromosomes paired at first metaphase:—

$$5A-5B = a \qquad 5A-5S = x$$
$$5A-5D = b \qquad 5B-5S = y$$
$$5B-5D = c \qquad 5D-5S = z$$

Values for a, b and c were observed and can be taken from column (4) of table 1. These values may be accepted as the frequencies with which chiasma formation occurred between the long arms of the stipulated chromosomes. The only ambiguity in these observations stems from the inclusion of occurrences of triradial trivalents, recorded in column (3), since these involve a minimum of two chiasmata which might be distributed:—(i) one between the telocentrics and one between a telocentric and a complete chromosome or (ii) both between a telocentric and the complete chromosome. In the latter instance there would have been no direct association between the telocentrics. However, the absolute frequency of these trivalents was so low that little error is introduced by accepting all as due to distributions involving direct chiasma formation between the telocentrics. With this proviso, however, the relative affinities of chromosomes 5A, 5B and 5D may be taken from table 1 column (4), where they are expressed as the percentages of cells in which the telocentrics paired, so that:—

$$a = 8$$
$$b = 8$$
$$c = 52.$$

These results are the first direct cytological observations—apparently in any organism—that have permitted an expression of differential affinity in quantitative terms.

The values for x, y and z cannot be obtained from direct observation since it was not possible to mark chromosome 5S of *Ae. speltoides*. However, the values can be estimated using the observed frequency with which a marked telocentric chromosome paired with an unmarked chromosome in the three types of hybrid with two group 5 chromosomes marked. Thus, for example, where chromosomes 5A and 5B were marked, the pairing of a marked with an unmarked chromosome could be of four types—5A with 5D, 5A with 5S, 5B with 5D, or 5B with 5S. Therefore, if the frequency of pairing between marked and unmarked chromosomes is called T, then

$$T_{(5A/5B)} = b+x+c+y.$$
Similarly, $T_{(5A/5D)} = a+x+c+z,$
and $T_{(5B/5D)} = a+y+b+z.$

The value for T in each formula can be obtained from table 1 by the summation of the frequencies of marked and unmarked pairing, recorded in columns (3), (8) and (11) $\times 2$. The frequencies for column (11) are doubled since each cell recorded had two instances of chiasma formation between marked and unmarked chromosomes. Thus,

$$T = \text{columns } (3)+(8)+(11)\times 2,$$

and $T_{(5A/5B)} = 89$
$$T_{(5A/5D)} = 81$$
$$T_{(5B/5B)} = 46.$$

Since values for a, b and c were derived by direct observation, they can be substituted in the above formulae, as can the values for T. When this is done,

$$89 = 8+x+52+y$$
$$81 = 8+x+52+z$$
$$46 = 8+y+8+z$$

and
$$x+y = 29$$
$$x+z = 21$$
$$y+z = 30.$$

Solving for x, y and z shows that,

$$x = 10$$
$$y = 19$$
$$z = 11.$$

Although these estimates are likely to be subject to some error, it is probably not large; and the overall determinations of affinities must at least indicate their order, and in general terms, the degree of expression. Consequently it can be unquestionably asserted that there are pronounced differences between the relative affinities of the four chromosomes for each other, which from the present evidence can be expressed as follows:—

$$5A-5B = 8$$
$$5A-5D = 8$$
$$5B-5D = 52$$
$$5A-5S = 10$$
$$5B-5S = 19$$
$$5D-5S = 11.$$

This result may be interpreted as representing the relative similarity or difference of the chromosomes of homoeologous group 5 in structure and gene content. However, this is not the only interpretation, and the meaning of these observations will be considered more extensively later.

Nevertheless, it seemed useful to attempt to compare the affinity of homoeologues with that of fully homoeologous chromosome, under comparable genetic conditions. Unfortunately such a comparison could not be made on a group 5 chromosome because appropriate parental material was not available. Instead use was made of a parental stock of *T. aestivum* Chinese Spring that had 20 normal pairs of chromosomes but in which chromosome 6B was represented by a telocentric for one arm only, in the trisomic condition. A plant of this type was pollinated by *Ae. speltoides* and 29-chromosome hybrids produced, in which the telocentric of 6B was present in double dose. The pairing at first metaphase of meiosis of these two fully homologous telocentrics was scored in the same way as that already described for homoeologous telocentrics (table 2).

The two 6B telocentrics paired together and formed a bivalent in the majority of cells, and only rarely did they pair with an unmarked chromosome—presumably a homoeologue. Applying the same system of measurement as to the group 5 homoeologues, the numerical estimate of the affinity of the homologous 6B telocentrics for each other is 86. In these terms, therefore, chromosomes 5B and 5D have 6o

TABLE 2

Pairing behaviour of two homologous 6B telocentric chromosomes at MI of meiosis in
T. aestivum × Ae. speltoides *hybrids*

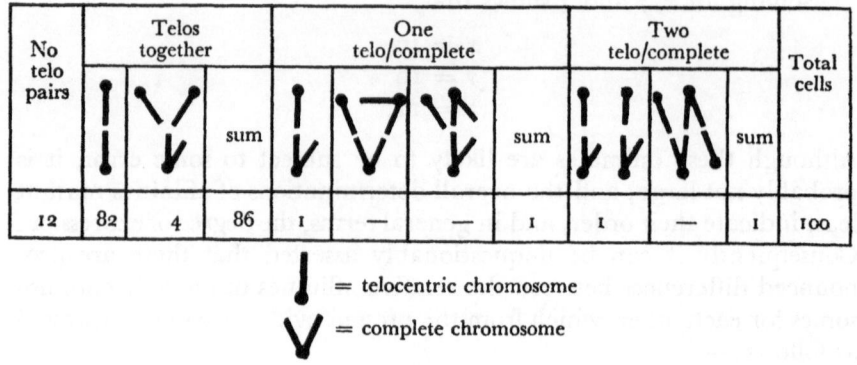

No telo pairs	Telos together		sum	One telo/complete			sum	Two telo/complete		sum	Total cells
12	82	4	86	I	—	—	I	I	—	I	100

= telocentric chromosome

= complete chromosome

per cent. of the affinity for each other displayed by the 6B telocentrics. Yet in normal euploid individuals of *T. aestivum* the 6B homologues almost always pair at meiosis, while the 5B and 5D homoeologues never do so. This nicely demonstrates the narrow margin of distinction between homologues and homoeologues upon which the discriminatory pairing imposed by the 5B system normally operates.

6. THE INTERPRETATION OF DIFFERENTIAL AFFINITY

The present observations, which uniquely demonstrate the validity of the concept of differential affinity originally proposed by Darlington (1928, 1937), permit the phenomenon to be described numerically. In so doing they force the acknowledgment of a problem which, though long recognised (Darlington, 1937; Dobzhansky, 1941), has been largely ignored; namely, the extent to which the capacity of chromosomes to pair at meiosis measures similarities in their overall structures and in the totality of their genetic equivalence. That the extent of meiotic pairing is not a strict measure of such similarity is shown by the effect of chromosome 5B on the meiosis of *T. aestivum*. Judgments made from the meiosis of material displaying this effect would ascribe no equivalence to homoeologues, despite the cytological similarities revealed in its absence and the genetic similarities displayed by nulli-somic-tetrasomic compensation tests. It seems unlikely that further genetic restrictions of pairing specificity, applying to the entire chromosome complement, occurred in the present hybrid situations. Nevertheless, other processes may have influenced the pairing of particular chromosomes in a way that rendered the frequency of pairing a distorted measure of gross genic and structural correspondence.

However, disregarding these possibilities in the first instance, if the group 5 pairing is assumed to measure total genetic congruence, then the closest relationship exists between chromosomes 5B and 5D, followed at a considerable distance by that between 5B and 5S. Indeed, using the evidence of the behaviour of the 6B telocentrics, about 60 per cent. of the genetic equivalence of completely homozygous homologues could be ascribed to the 5B and 5D homoeologues. An interpretation of this type, accepting the assumption of a common origin of homoeologues, implies that the differential affinity of chromosomes is an expression of their relative evolutionary proximities. Consequently chromosome 5A must be considered to have diverged from 5B and 5D to a much greater extent than 5B and 5D have from each other. If true, this presents to our view a new facet of the cytogenetic structure of *T. aestivum*, indicating that chromosomes in the A and B genomes—that have been together longest, first in tetraploid then in hexaploid wheat—have diverged from each other widely. This is in accord with other evidence, primarily from the study of mutagenesis at different levels of polyploidy, showing that the tetraploid species of *Triticum*, with only the A and B genomes, have much less duplication of genetic material than the hexaploid. On the present evidence, the divergence must, so far as group 5 is concerned, have resulted principally from changes in chromosome 5A, since the original similarity, which may be presumed for all homoeologues, is still retained to any considerable extent only by 5B and 5D. Whether the proposals, derived from the study of the group 5 situation, are of general application to the majority of chromosomes in each genome can only be determined by similar work on other homoeologous groups. However, if the same patterns of relationships are usual, then their interpretation in terms of genomic divergence and the overall genetic distinctions of chromosomes must be valid.

There are at present, however, certain anomalies that qualify the acceptance of determinations of differential affinity solely as measurements of gross genetic equivalence. First, although the B genome of *T. aestivum* was almost certainly derived from *Ae. speltoides*, or a close relative (Sarkar and Stebbins, 1956; Sears, 1956; Riley, Unrau and Chapman, 1958), chromosomes 5B and 5S—while displaying the second-highest affinity—paired considerably less frequently than 5B and 5D. Since the acceptance of an interpretation of the results in terms of gross similarities requires the assumption that 5B is little changed, this must imply that 5S has undergone marked changes since it gave rise to 5B. While this is possible, it contrasts with the conditions of the chromosomes of the A and D genomes of the hexaploid species relative to those of the corresponding genomes of the diploid species, from which they were derived. Corresponding chromosomes of these diploids and the hexaploid can still be described as homologous, since they are able to pair at meiosis even when the restriction imposed by 5B is operative. The present evidence seems to indicate that, even if *Ae. speltoides* is the B genome donor, its hybrid with *T. aestivum* would show little or no meiotic pairing were the 5B activity not suppressed. For, by analogy with the behaviour of chromosomes of the A and D genomes, chromosomes 5B and 5S should be regarded as homologous, yet they have less affinity than some homoeologues. Some doubt must remain, therefore, that the frequency of 5B-5S pairing is even an approximate indication of their overall genetic relationship.

Similar doubt arises from the high pairing between chromosomes 5B and 5D since this result leads to the expectation that there would be good compensation in nullisomic-tetrasomic combinations in which these chromosomes were in altered dosage. By contrast less satisfactory compensation would be expected when chromosome 5A was in altered dosage in the same nullisomic-tetrasomic genotypes as either of the other two group 5 chromosomes. However, such expectations were not realised in Sears' (1965) studies, since although there was a slight superiority of the nullisomic 5D tetrasomic 5B condition over all other group 5 combinations, this was not repeated in nullisomic 5B tetrasomic 5D. While it may be that differential affinity is simply a more sensitive measure of overall genetic correspondence than is compensatory capacity in nullisomic-tetrasomic conditions, the alternative possibility must also be recognised that different properties may be measured by the two methods. If different aspects of chromosomal relationships are indeed involved, then the compensation test—which depends upon the occurrence of similar genetic activities in homoeologues—is more likely to measure gross genetic correspondence.

By what alternative conditions therefore might the differential affinity of homoeologues be determined? It is only possible to speculate upon other conditions but, for example, synapsis might be visualised as dependent upon contacts made between particular regions of chromosomes that are of limited extent and number. If this were so, differential affinity might measure not the overall similarities of chromosomes but the relative structural and genetic congruence of the contact regions. Homoeologues with contact regions that had undergone evolutionary divergence would then pair infrequently despite an occurrence of general correspondence in their genetic activities.

An alternative to this notion of limited but autonomously functioning contact points is the postulate that the pairing specificity of each chromosome is independently determined by gene activity. Pairing would thus result in part from the structural equivalence of the potential pairing partners but it would also be subject to variation caused by the presence of particular gene products. In terms of the expression of differential affinity there would be no detectable difference between the outcome of the operation of either specific pairing genes or limited contact regions. However, in the investigation of homoeologous pairing between the chromosomes of *T. aestivum*, it will probably be easier to test the ideas concerning restricted contact regions. This arises from the largely distal chiasma formation, indicating that pairing is initiated distally and that the hypothetical contact regions may be supposed to be located towards the ends of wheat chromosomes. Consequently homoeologues with structural modifications confined to their terminal regions could be used to test whether differential affinity is determined by particular regions of the potential partners.

Despite the scepticism expressed regarding the interpretation of differential affinity in terms of gross genetic equivalence, the possibility nevertheless remains that its measurement for the chromosomes of group 5 may provide a true estimate of the strength of genetic homoeology. In any event, these observations certainly indicate the direction in which the study of the genetic structure of wheat should progress. Whether obtained by cytological or by genetical procedures, quantitative expressions of the relationships of homoeologous chromosomes to each other will illuminate the contemporary genetic status of *T. aestivum* as well as reveal its evolutionary history and some of its future potentialities in greater detail.

7. SUMMARY

1. Hybrids between common wheat, *Triticum aestivum* (2n = 6x = 42), and *Aegilops speltoides* (2n = 14) have high levels of chromosome pairing at meiosis, with numerous multivalents. The assumption that this pairing was between homoeologous chromosomes was tested by producing *T. aestivum* × *Ae. speltoides* hybrids in which two wheat chromosomes were marked by a telocentric condition. When the marked chromosomes were not homoeologous they never participated in the same configuration. By contrast, when the marked chromosomes were homoeologues, belonging to group 5, they were frequently involved in the same configurations as bivalents, trivalents or quadrivalents. Thus pairing in these hybrids was shown to be homoeologous, and the genotype of *Ae. speltoides* must therefore suppress the activity of chromosome 5B of wheat by which homoeologous pairing is normally inhibited.

2. Since the telocentric of only one arm of chromosomes 5A, 5B and 5D was used, and yet each was capable of pairing with either of the others *directly as a bivalent*, the chromosome arms represented are shown to correspond.

3. Comparisons of the relative frequencies with which the four chromosomes—5A, 5B and 5D from *T. aestivum* and 5S from *Ae. speltoides*—paired together showed pronounced differential affinities of these chromosomes for each other. The association of 5B with 5D was much more frequent than any of the other five possible two-by-two

E

associations. The next most common pairing was between chromosomes 5B and 5S. A quantitative estimate of the differential affinity of particular chromosomes at meiosis was thus made, apparently for the first time in any organism. The interpretation of this differential affinity is discussed in relation to its possible indication of the gross genetic equivalence of the participating chromosomes.

8. REFERENCES

DARLINGTON, C. D. 1928. Studies in *Prunus*, I and II. *J. Genet., 19*, 213-256.

DARLINGTON, C. D. 1937. *Recent Advances in Cytology*, 2nd Edition. Churchill, London.

DOBZHANSKY, T. 1941. *Genetics and the Origin of Species*, 2nd Edition. Columbia University Press, New York.

MCFADDEN, E. S., AND SEARS, E. R. 1946. The origin of *Triticum spelta* and its free-threshing hexaploid relatives. *J. Hered., 37*, 81-89, 107-116.

RILEY, R. 1960. The diploidisation of polyploid wheat. *Heredity, 15*, 407-429.

RILEY, R. 1965. The genetic regulation of meiotic behaviour in wheat and its relatives. *Proc. 2nd Int. Wheat Genet. Symp.* (In the press.)

RILEY, R., AND CHAPMAN, V. 1958. Genetic control of the cytologically diploid behaviour of hexaploid wheat. *Nature, Lond., 182*, 713-715.

RILEY, R., AND CHAPMAN, V. 1964a. Cytological determination of the homoeology of chromosomes of *Triticum aestivum*. *Nature, Lond., 203*, 156-158.

RILEY, R., AND CHAPMAN, V. 1964b. The effect of the deficiency of the long arm of chromosome 5B on meiotic pairing in *Triticum aestivum*. *Wheat Inf. Service, 17-18*, 12-15.

RILEY, R., AND KEMPANNA, C. 1963. The homoeologous nature of the non-homologous meiotic pairing in *Triticum aestivum* deficient for chromosome V(5B). *Heredity, 18*, 287-306.

RILEY, R., KIMBER, G., AND CHAPMAN, V. 1961. The origin of the genetic control of the diploid-like meiotic behaviour of polyploid wheat. *J. Hered. 52*, 22-26.

RILEY R. AND LAW C. N. 1965. Genetic variation in chromosome pairing. *Adv. Genetics, 13*, 57-114.

RILEY, R., UNRAU, J., AND CHAPMAN, V. 1958. Evidence on the origin of the B genome of wheat. *J. Hered., 49*, 91-99.

SARKAR, P., AND STEBBINS, G. L. 1956. Morphological evidence, concerning the origin of the B genome of wheat. *Amer. J. Bot., 43*, 297-304.

SEARS, E. R. 1954. The aneuploids of common wheat. *Res. Bull. Mo. agric. Exp. Stn.* 572.

SEARS, E. R. 1956. The B genome in wheat. *Wheat Inf. Service, 4*, 8-10.

SEARS, E. R. 1966. Nullisomic-tetrasomic combinations in hexaploid wheat. In *Chromosome Manipulations and Plant Genetics* (ed. R. Riley and K. R. Lewis) a supplement to *Heredity, 20*, 29-45. Oliver and Boyd, Edinburgh.

ZOHARY, D., AND IMBER, D. 1963. Genetic dimorphism in fruit type in *Aegilops speltoides*. *Heredity, 18*, 223-231.

BIOMETRICAL ANALYSIS USING CHROMOSOME SUBSTITUTIONS WITHIN A SPECIES

C. N. LAW

Plant Breeding Institute, Cambridge, England

1. INTRODUCTION

THE development of appropriate aneuploid material in a polyploid species enables the substitution of single chromosomes from one variety into another. In this way, chromosomes from a population may be " assayed ", one at a time, in a common and constant genetic background. The simplification which thereby occurs as a result of replacing the gene by the chromosome as the unit of inheritance, provides a favourable situation for studying the genetics of quantitative characters. The reduction of many determinants to a comparative few allows a level of sophisticated technique to be introduced into the study of polyploid organisms—a level that up to now has been largely confined to the study of *Drosophila melanogaster*.

Some of the manipulative procedures which can be applied to this kind of genetic material will be described in this account. The outcome of such techniques, it may be hoped, will provide information at three levels. At the level of application it may pin-point, so far as the economically useful polyploids are concerned, a localisation of genetic effect which has been and will be of value to those interested in crop improvement. Secondly, the genetic architecture, which may be revealed, will provide impetus to studies on the origin and evolutionary development of polyploids, particularly if it is possible to relate this architecture to the ancestral type; and thirdly, these procedures will provide a ready means of checking the predictions and genetic estimates made by the use of more analytical approaches to quantitative inheritance.

At present, the requisite aneuploid material is available in only one polyploid organism, the hexaploid wheat, *Triticum aestivum* ($2n = 6x = 42$), although it may be supposed that other polyploids will soon be in a similar position. In *T. aestivum*, Sears (1954) was the first to produce the complete series of 21 monosomics and 21 nullisomics using the variety Chinese Spring. These have been, and continue to be, of inestimable value to those engaged in the cytogenetical manipulation of chromosomes within the wheat genotype. Sears (1953) has also described the cytological techniques used in exploiting aneuploid material in the production of inter-varietal substitutions. These techniques have been further elaborated by Unrau, Person and Kuspira (1956). The value of the technique from the point of view of a chromosome assay has also been demonstrated most decisively by Kuspira and Unrau (1957), who showed that appreciable effects on yield resulted from the substitution of single pairs of chromosomes of the varieties Hope, Thatcher and Timstein for their homologues in Chinese Spring. It now remains to relate such straightforward assay techniques to the more elaborate and informative procedures used in biometrical genetics. The following account goes some way to describing the scope of inter-varietal substitutions in the terms of biometry.

2. INTER-VARIETAL SUBSTITUTIONS AND THE STUDY OF QUANTITATIVE INHERITANCE

Three types of genetic parameter may be used to describe the genotype of whole chromosome substitution lines and the hybrids derived from them. These refer to the additive effects of each chromosome, the departures from additivity which are due to within-chromosome interaction, and the departures from additivity which arise as a result of between-chromosome interaction. The symbols used by Hayman and Mather (1955) to characterise a digenic situation, d for additive effects, h for dominance, and l, j and i for non-allelic interaction, may be used in the genetic description of substitution lines. In this case, however, it must be noted that the symbols refer to whole chromosomes instead of to single genes, so that the symbol h, referring to dominance or interaction within a locus for the digenic situation proposed by Hayman and Mather, is in the case of substitution lines referable to interaction within an entire chromosome. Both within-locus and between-locus interactions are consequently combined in this parameter. Similarly, d, for the additive effect of a whole chromosome may conceal interactions of an additive $\times$ additive kind (i). The simplifying assumption that only dichromosomal interaction occurs has also been made.

For ease of description a simplified model of a substitution series may be employed in which there are only three homologous pairs of chromosomes. The chromosomes of the donor variety may be represented by capital letters, AA BB CC, and those of the recipient variety by the lower-case, aa bb cc. With three pairs of chromosomes, only three substitutions of donor chromosomes into the recipient variety are possible, so that altogether five genotypes are available for study and manipulation, AA BB CC, AA bb cc, aa BB cc, aa bb CC, and aa bb cc. The expressions of these genotypes may be formulated as follows, where m refers to the general mean from which all the parameters are defined,

$$
\begin{array}{ll}
\text{AA BB CC} & m+d_a+d_b+d_c+i_{ab}+i_{ac}+i_{bc} \\
\text{AA bb cc} & m+d_a-d_b-d_c-i_{ab}-i_{ac}+i_{bc} \\
\text{aa BB cc} & m-d_a+d_b-d_c-i_{ab}+i_{ac}-i_{bc} \\
\text{aa bb CC} & m-d_a-d_b+d_c+i_{ab}-i_{ac}-i_{bc} \\
\text{aa bb cc} & m-d_a-d_b-d_c+i_{ab}+i_{ac}+i_{bc}
\end{array}
$$

It is clear, because there are only five observations to solve seven unknowns, that even with a simplified model such as this, a complete solution of the genetic equations is not possible. A number of manipulative procedures can, however, be applied to this type of material, which will allow some separation of the genetic parameters. Three of these will be described below.

(i) Method I. The substituted series along with the donor and recipient varieties

Using the simplified substitution series again, an estimate of the additive contribution of each chromosome may be obtained from comparisons of the type $\frac{1}{2}$ (AA bb cc—aa bb cc), which for the parameters given above will equal $d_a - i_{ab} - i_{ac}$, in other words, the additive contribution of chromosome A—a, minus the summed interactions of A—a with the other two chromosomes. A similar result will occur for chromosome B—b and C—c. Now, if all these three additive estimates, from A—a, B—b and C—c are combined, then an estimate will be derived which will be equivalent to the sum of all the d effects minus twice the sum of all the i effects, that is $A - 2I$, where $A = d_a + d_b + d_c$ and $I = i_{ab} + i_{ac} + i_{bc}$. Still further, the comparison $\frac{1}{2}$ (AA BB CC—aa bb cc) will give an estimate which will be equal to $d_a + d_b + d_c$ or A.

A method is therefore available which enables the summation of all the additive contributions of each chromosome to be separated from twice the summation of all the between-chromosome interactions. The two parameters, A and I, can consequently be estimated and a general test for between-chromosome interactions achieved.

Although the relatively rapid test which can be achieved by this means is an advantage, the possible weaknesses and inefficiencies of this method are numerous. This will be particularly apparent when the localisation of between-chromosome interactions within the genotype is required. Furthermore, the balancing of + and − additive and interactive effects, to which the summation of a number of parameters must lead, may at times prove to be rather misleading. Thus, a non-significant A or I estimate may not always be indicative of their negligible effects. To avoid these complications other methods of chromosome manipulation are available and may be attempted.

(ii) Method II. The diallel cross

It is possible to cross, in a diallel fashion, two substitution lines and the recipient variety and obtain an estimate of between-chromosome interactions involving only two pairs of chromosomes. The crossing scheme for the simplified genotype with three homologous chromosomes is illustrated in table 1.

Obviously a great number of parameters are involved and only six observations are available, so that as in Method I a complete estimation cannot be achieved. However, an estimate of between-chromosome interaction can be obtained by selecting four genotypes from this crossing procedure, Aa Bb cc, Aa bb cc, aa Bb cc and aa bb cc. By adding the first and the last of these genotypes and subtracting the middle two, both the additive and within-chromosome interaction effects will be removed and in terms of parameters, $l_{ab}+j_{a/b}+j_{b/a}+i_{ab}$ will be left as the measure of between-chromosome interactions. The presence of interaction between the chromosomes A—a and B—b can, therefore, be detected by this comparison. The estimate, however, because it involves four parameters is still difficult to interpret. Thus, the significant interactions detected may result from the interaction of

TABLE I

The diallel cross

♀ \ ♂	AA bb cc	aa BB cc	aa bb cc
aa bb cc	Aa bb cc $m+h_a-d_b-d_c-j_{b/a}-j_{c/a}+i_{bc}$	aa Bb cc $m-d_a+h_b-d_c-j_{a/b}+i_{ac}-j_{c/b}$	aa bb cc $m-d_a-d_b-d_c+i_{ab}+i_{ac}+i_{bc}$
aa BB cc	Aa Bb cc $m+h_a+h_b-d_c+l_{ab}-j_{c/a}-j_{c/b}$	aa BB cc $m-d_a+d_b-d_c-i_{ab}+i_{ac}-i_{bc}$	—
AA bb cc	AA bb cc $m+d_a-d_b-d_c-i_{ab}-i_{ac}+i_{bc}$	—	—

either the two heterozygous chromosomes, or the two homozygous chromosomes, or one or other of the two homozygous-heterozygous combinations, or, indeed, from all, or some combination of these interactions; there is no means of telling which may be responsible. However, the fact that only four parameters are involved suggests that a balancing of positive and negative interactive effects is less likely to influence the detection of a between-chromosome interaction than is the case in Method I, in which a large number of interactive parameters may be summed.

Estimates of other genetic parameters may also be achieved from the diallel cross method and these refer to the study of within-chromosome effects. Thus, an estimate of within-chromosome interaction may be derived for chromosome A—a, for example, from the comparison of $\frac{1}{2}(-\text{AA bb cc}+2$ Aa bb cc—aa bb cc), which provides $\hat{h}_a = h_a-j_{b/a}-j_{c/a}$ as the estimate of within-chromosome interaction. The estimates of h_a and h_b are consequently only completely satisfactory when between-chromosome interactions are zero. When the latter are present, then the estimate of h_a ($\hat{h}_a$) and h_b ($\hat{h}_b$) will be negatively correlated with the estimates of between-chromosome interaction.

A similar situation arises when the additive effects of each chromosome are estimated. As in the case of Method I, the best estimate is derived from the comparison of $\frac{1}{2}$ (AA bb cc—aa bb cc), which is equal to $d_a - i_{ab} - i_{ac}$. Once again, therefore, the estimates of d_a and d_b will only be accurate when between-chromosome interactions are not involved. When this occurs, then both the within-chromosome interaction and additive estimates will involve components of the between-chromosome interaction estimates. Indeed, the difficulties of deriving a unique estimate of these two types of parameter in the presence of non-allelic interaction have been remarked upon by Hayman (1958).

(iii) Method III. Reciprocal substitutions

So far, the methods which have been described concern the manipulation of a single series in which the chromosomes of one variety are substituted into another. It is also possible, if the requisite aneuploid material is present in two varieties, to produce reciprocal substitutions, so that not only are single chromosomes substituted from the one variety into the other, but also vice versa. The outcome of such a procedure is a reciprocally substituted series and is perhaps the nearest to an orthogonal arrangement of wheat chromosomes which can be achieved at the moment. Thus, to use a simplified method of description again, where A—a stands for the chromosomes which are being substituted and B—b refers to the two genetic backgrounds, then reciprocal substitutions for the A—a chromosome may be depicted as AA bb and aa BB, and the two varieties as AA BB and aa bb respectively. An orthogonal arrangement is produced therefore in which the substituted chromosomes can be tested for additivity and interaction with the remaining twenty chromosomes of the background.

$$
\begin{array}{ll}
\text{AA BB} & m+d_a+d_b+i_{ab} \\
\text{AA bb} & m+d_a-d_b-i_{ab} \\
\text{aa BB} & m-d_a+d_b-i_{ab} \\
\text{aa bb} & m-d_a-d_b+i_{ab}
\end{array}
$$

Four parameters, m the general mean, d_a the additive effect of chromosome A—a, d_b, the additive effect of the genetic background B—b, and i_{ab}, the interaction of chromosome A—a with the genetic background, require estimation. Four observations can be made and are such that it is possible to obtain exact estimates of all the parameters available. A similar procedure can be carried out for any substituted chromosome, so that the behaviour of each chromosome with respect to the genetic background of the two varieties involved can be obtained.

Furthermore, the complications found in the estimation of d and h in the diallel cross are to a certain extent removed by the use of reciprocal substitutions. Thus, exact estimates, with respect to the two homozygous genetic backgrounds, B and b, can be achieved from material of this nature. The estimates of d_a and d_b follow simply from the observation and expectations for the four genotypes mentioned earlier, whereas h_a can be derived from the comparison:—

$$\tfrac{1}{4}\,(-\text{AA BB}+2\ \text{Aa BB}-\text{aa BB}-\text{AA bb}+2\ \text{Aa bb}-\text{aa bb}) = h_a,$$

and h_b from the comparison:—

$$\tfrac{1}{4}\,(-\text{AA BB}+2\ \text{AA Bb}-\text{AA bb}-\text{aa BB}+2\ \text{aa Bb}-\text{aa bb}) = h_b.$$

The disadvantage inherent in the diallel cross method can consequently be removed and to some degree the contributions of additive and within-chromosome interaction effects to the phenotype studied.

3. THE DIALLEL CROSS USING THE HOPE SUBSTITUTIONS

The substitution series developed by E. R. Sears, in which the chromosomes of the variety Hope are separately substituted for their homologues in Chinese Spring, has been used to investigate the suitability of two of the techniques described. The first experiment using this material employed the diallel crossing procedure.

It is almost impossible to cross all the 21 substitution lines of Hope along with the reciprocal variety, Chinese Spring, in the diallel fashion proposed. A restricted chromosome sample has, therefore, been used. Now, the classification of the hexaploid wheat chromosomes into the three genomes, A, B, and D and also into the seven homoeologous groups (Sears, 1954) provides a "skeletal" genetic architecture of wheat. Accordingly, any restriction in the chromosome sample size must, if the maximum amount of genetic information is to be derived, take such a "skeletal" architecture into account. The chromosomes sampled must as a consequence be chosen so that the distribution of genetic effects amongst the genomes and homoeologous groups can be isolated and comparisons made between them. The experiment was therefore carried out using Chinese Spring and nine Hope substitution lines, in which, in turn, chromosomes 1A, 1B, 1D, 3A, 3B, 3D, 7A, 7B and 7D of Hope were separately substituted into Chinese Spring. Among the lines used there were therefore three involving the substitutions of chromosomes in each of the three genomes and also the three chromosomes belonging to each of the three homoeologous groups, 1, 3 and 7.

The F_1 plants produced in the 10 × 10 diallel cross were grown out in the field as a randomised block experiment in 1962; five blocks were set out at seven plants per plot in each block, the plants being spaced six inches apart within plots.

Four characters were studied, yield, ear number, 500 grain weight and height. The estimates for the additive effect of each chromosome (d), within-chromosome interactions (h) and between-chromosome interactions ($l+2_j+i$) for each of these four characters are given in tables 2-6.

TABLE 2

Estimates of additivity and within-chromosome interaction for yield per plant, ear number per plant, 500 grain weight and height per plant

Estimates of additive chromosome effects (d)

	1A	3A	7A	1B	3B	7B	1D	3D	7D	
Yield	−0·64***	+0·38*	−0·17	−0·25	−0·47*	−0·23	+0·58**	−0·25	−0·29	−1·34
Ear Number	+0·10	+0·46***	−0·18	−0·54***	−0·23	−0·12	+0·02	−0·29*	−0·18	−0·96
Grain Weight	+0·53***	−0·20*	+1·01***	+0·71***	0·00	−0·27**	+0·71***	−0·36***	−0·32**	+1·81
Height	−2·05***	−1·37**	+0·79	−0·31	+0·41	−0·25	−0·31	+0·09	+0·66	−2·34

Estimates of within-chromosome interaction (h)

	1A	3A	7A	1B	3B	7B	1D	3D	7D	
Yield	+0·38	−0·27	+0·74	+0·40	−1·23**	−1·14**	+0·54	+0·82	−0·75	−0·51
Ear Number	−0·08	−0·60*	+0·38	+0·18	−0·61*	−0·69**	+0·20	+0·29	−1·05***	−1·98
Grain Weight	+0·32	−0·28	−0·01	+0·69***	+0·13	+0·22	−0·07	+0·09	+0·31	+1·40
Height	+1·12	−1·18	+1·67	+2·30*	−2·22*	−3·03**	−0·91	+1·13	−0·12	−1·24

* P 0·05–0·01 ** P 0·01–0·001 *** P<0·001

TABLE 3

Between-chromosome interaction for yield per plant in grammes

	1A	3A	7A	1B	3B	7B	1D	3D	7D
7D	+1·58*	+1·88*	+1·37*	+1·60*	+1·50*	+3·79***	−0·36	+0·62	+11·98
3D	−0·13	+0·09	−2·08**	−0·55	−0·75	+1·17	−2·70***		−4·33
1D	−0·97	−0·98*	−0·06	+0·61	−0·01	+0·13			−4·34
7B	−0·17	+2·51***	+0·98	+1·34	+1·98**				+11·73
3B	+0·07	+0·39	+0·99	+0·64					+4·81
1B	−0·51	+1·19	−0·81						+3·51
7A	−1·09	−0·43							−1·13
3A	+0·63								+5·28
1A									−0·59

Total +26·92

* P 0·05–0·01   ** P 0·01–0·001   ***P<0·001

TABLE 4

Between-chromosome interaction estimates for the number of ears per plant

	1A	3A	7A	1B	3B	7B	1D	3D	7D
7D	+0.88*	+1·54***	+0·79*	+1·35***	+1·26**	+2·10***	+0·12	+0·45	+8·49
3D	−0·23	+0·26	−1·18**	+0·58	−0·57	+0·59	−1·52***		−1·62
1D	−0·51	−0·30	−0·48	+0·29	−0·05	+0·19			−2·26
7B	−0·14	+1·53***	+0·12	+0·65	+0·91*				+5·95
3B	−0·38	+0·68	+0·35	+0·56					+2·76
1B	−0·27	+0·92*	−0·64						+3·44
7A	−0·74	−0·41							−2·19
3A	+0·32								+4·54
1A									−1·07

Total +9·02

* P 0·05–0·01   ** P 0·01–0·001   *** P <0·001

TABLE 5

Between-chromosome interaction estimates for height per plant

	1A	3A	7A	1B	3B	7B	1D	3D	7D	
7D	+1·66	+0·34	−0·61	+1·94	−0·02	+3·68*	+1·27	+0·54		+8·80
3D	+0·96	+0·92	−5·02**	+0·38	+0·50	+4·12*	+0·12			+2·52
1D	+0·39	+0·59	−0·61	+0·48	+5·59**	+3·00				+10·83
7B	+1·83	+4·36*	+1·68	+3·67*	+5·95***					+28·29
3B	−0·86	+2·63	+1·61	+2·59						+17·99
1B	−0·30	+1·47	−2·39							+7·84
7A	−2·11	−0·64								−8·09
3A	+0·33									+10·00
1A										+1·90

Total +40·04

* P 0·05-0·01   ** P 0·01-0·001   *** P<0·001

TABLE 6

Between-chromosome interaction estimates for 500 grain weight

	1A	3A	7A	1B	3B	7B	1D	3D	7D	
7D	−0·15	+0·07	−0·30	−0·17	+0·06	+0·02	+0·03	+0·39		−0·05
3D	−0·42	+0·69*	−0·37	−0·53	−0·19	+0·12	−0·19			−0·50
1D	−0·29	−0·06	−0·02	−0·35	−0·37	−0·28				−1·53
7B	−0·33	+1·00**	−0·11	−0·71*	+0·49					+0·20
3B	−0·44	+0·30	+0·12	−0·32						−0·35
1B	−1·19***	+0·10	−1·00**							−4·17
7A	−0·58	+0·36								−1·90
3A	+0·25									+2·71
1A										−3·15

Total −4·37

* P 0·05-0·01   ** P 0·01-0·001   *** P<0·001

It is evident from these tables that significant genetic effects are found both within and between chromosomes. Furthermore, for three of the characters yield, ear number and height—the frequent occurrence of between-chromosome interactions suggests that the effects of a chromosome in one genetic background are unlikely to be reproduced in exactly the same manner and extent in other genetic backgrounds. Thus, the substitution of a chromosome from one variety may have appreciable effects on the yield of one recipient variety, but have little effect at all on a second recipient. It is unlikely, therefore, that an assay carried out on a chromosome substituted into an economically unadapted variety such as Chinese Spring will provide any exact evidence as to its behaviour in more adapted and economically useful genetic backgrounds. This is not to say that useful information may not accumulate from the consideration of chromosome assays by the technique of inter-varietal substitution. It is quite probable that the beneficial response obtained by substituting a particular chromosome into an adapted background will also be shown, although not necessarily to the same extent, in other adapted genetic backgrounds. A consistency of behaviour may thus result which will be of extreme value to the plant breeder.

At the moment, however, the wide occurrence of between-chromosome interaction suggests that the most profitable line of advance, until such adapted material is available, is to record the incidence, magnitude and position of the various genetic constants which have been estimated. Such records must be viewed in relation to the known skeletal genetic architecture of genomes and homoeologous groups of wheat, in the hope that further details of the genetic structure will emerge. The pattern of genetic behaviour which may result from this line of advance may then enable the plant breeder to restrict his interest to some part or parts of the wheat genotype.

Accordingly the Hope substitution diallel has been considered from this standpoint and the distribution of significant genetic effects to the higher order classification of genomes and homoeologous groups is presented here. Similarly, the differences which may result between the various characters studied must not be overlooked. Such differences may indeed be of importance, since they will reflect the different past and present selective pressures which have been applied to these characters and consequently the future prospects of selection will be better understood.

(i) Genetic balance

The sign of the genetic effects which have been estimated is of interest. Both + and − estimates occur in all the three main groups of parameters, additive effects, and within- and between-chromosome interactions. The distribution of significant + and− effects to these three groups is, however, not the same. For example, for the characters yield, ear number and height, significant negative estimates are more frequent than positive estimates for the additive and within-chromosome interactions effects. Between-chromosome interactions are, however, the reverse and positive interactions are the most frequent. This is shown even more by the totals obtained by summing all the additive, all the within-chromosome interaction, and all the between-chromosome interaction estimates. Whenever the summed estimates for within-chromosome effects are examined, negative totals are found; for between-chromosome interactions, however, the summed totals are all positive.

This pattern is departed from in one case only and this concerns the character grain weight. Here, + and − effects are equally distributed among the eight significant additive estimates, whereas for within-chromosome interactions only one estimate is significant and this is positive. Similarly, between-chromosome interactions show an almost equal distribution of positive and negative significant estimates. This character, therefore, behaves in a more ambidirectional manner with both + and − effects occurring with equal frequency among the estimates, irrespective of whether they refer to within- or between-chromosome effects.

It may be that the consistent pattern obtained for yield, ear number and height results from pleiotropy and the same genes are concerned in the control of all three characters. For ear number and yield this must be true, since ear number is obviously a major component of yield. Height, however, has not such an obvious connection with either yield or ear number and, indeed, although some similarities between the estimated parameters can be shown, both the sign and distribution of the estimates for this character are such as to suggest that, in part, different genetic systems are involved. Pleiotropy cannot, therefore, account entirely for the consistent effects observed for these characters. As a result, the preponderance of negative within-chromosome effects and the excess of positive between-chromosome effects for these characters is less likely to be due to a fortuitous association of negative and positive interacting genes to within- and between-chromosomes respectively. Another explanation has to be found.

Wheat is a cleistogamous inbreeder, so that adapted genotypes are presumably based on homozygous combinations of genes. On outcrossing, heterozygosity is introduced so that an unadapted complex of genes is formed. The expression of characters, therefore, which are of an adaptive nature, will decrease or increase depending on the direction in which previous selection, following rare hybridisation, has occurred. If selection has acted to increase the expression of a character, based on a homozygous balance, then outcrossing will produce a decline. By contrast, if selection has favoured a decrease, then disruption of the normal homozygous balance will act in a reverse manner. In each case, the disruption of genetic balance will run counter to the selection which has been responsible for its creation.

The negative interactions found within chromosomes for the three characters—yield, ear number and height—are features, therefore, which could be expected when an inbreeder is outcrossed. On the other hand, the large positive interactions for the same characters, but between chromosomes, are not explicable in terms of a disturbed homozygous genetic balance. An explanation of this may, perhaps, be obtained from examining the polyploid nature of wheat. Hexaploid wheat combines the chromosome complements of three diploid species with considerable genetic similarities. These three species are not, however, identical; as a result, the variation existing within a homoeologous group of chromosomes, in which the same loci but presumably different alleles are involved, may be regarded in much the same way as the variation between homologous chromosomes. In other words, allelic heterozygosity exists between non-homologous chromosomes. It is, however, a heterozygosity which is fixed and, indeed, it is this fixation of heterozygosity which may be one of the advantages of the polyploid species over related inbred diploids (Darlington, 1939; Lewis and John, 1963). The relationships between non-homologous chromosomes may consequently be regarded in a different manner from the relationships occurring between homologues. A degree of fixed hybridity exists between non-homologous chromosomes. Presumably, also, it is a degree of hybridity for which selection has been practised, so that a heterozygous balance occurs between non-homologous chromosomes. It follows, therefore, that the consequences of outcrossing on between-chromosome relationships are likely to be less drastic than on the relationships within chromosomes in which a completely homozygous balance is favoured. As a result, between-chromosome interactions which are counter to the direction of selection are less likely when outcrossing takes place.

For characters, such as yield and ear number, in which selection has been directional and for an increased expression, then, the negative within-chromosome interactions and the positive between-chromosome interactions which have been estimated follow exactly the directions which the genetic model has suggested. Height, also, follows a similar pattern, although it is doubtful whether such a character has been subjected to continuous directional selection. Thus, it is probably only recently that height has been subjected to any degree of directional selection and that for a reduction in height. Before this, selection has favoured perhaps the longest straw length which is commensurate with maintaining a good stand of wheat without too much lodging. As a character, therefore, height can be considered as being subjected to both stabilising and directional selection at times during the evolution of the hexaploid. If this is so, then the negative within-chromosome effects as opposed to the more prevalent positive between-chromosome effects may also be expected on outcrossing, although perhaps not to the same extent as in the case of a completely-directionally-selected character. Indeed, the occurrence of a significant positive within-chromosome interaction for height supports this view.

Of the characters studied, grain weight stands apart and has a predominantly additive inheritance. For the within- and between-chromosome interactions, when they occur, and also for the additive effects of each chromosome, both + and − effects are evenly distributed. Now, grain weight is probably a character to which an upper and lower limit is set biologically. On the one hand, a sufficiently large endosperm and embryo are necessary, if the individual organism is to have any chance of survival. On the other hand, too large an endosperm and embryo will, under natural conditions, have no particular advantage once the embryo is well established. Grain weight consequently exhibits all the features of a character subjected to stabilising rather than the directional selection proposed for the other three characters. The equal distribution of + and − interaction effects to both within- and between-chromosomes on outcrossing is consequently not surprising. To establish the type of genetic balance proposed earlier, directional selection which favours both within- and between-chromosome interactions must have occurred. With stabilising selection, neither + nor − interactions have any particular merit, so that not only will + and − interactions be equally distributed, but the presence of any interactions will be minimal (Breese and Mather, 1960). On outcrossing, the predominantly additive gene behaviour will still be maintained and the sign of the interactive effects will be as often positive as negative. This is exactly the result observed in this experiment.

The distribution of + and − effects among the genetic estimates gives a reasonable agreement with the model of genetic balance outlined. In barley, a similar result has been observed, although in this case it is the estimate of dominance which is relevant since chromosomal variation cannot be estimated. Thus, Johnson and Aksel (1959) have reported a correlation of recessive genes with high yield among 15 barley varieties used in a set of diallel crosses. In other words, there was a negative dominance effect on yield, which may well have arisen from a disruption of homozygous balance as a result of outcrossing. A similar conclusion follows also from the observation that hybrids between wheat varieties consistently show a reduction in meiotic stability compared with their parents (Riley and Law, 1965). A negative heterosis, or " outbreeding depression ", is consequently observed for certain meiotic characters in wheat and may well result from the disruption of a homozygous balance, between homologous chromosomes.

On the other hand, the demonstration that certain genes in barley appear to have a heterozygous advantage over the two homozygotes (Jain and Allard, 1960; Allard and Jain, 1962) indicates that the hypothesis of a homozygous genetic balance for inbreeding species must be considered with caution. Indeed, in the experiment described, this caution is emphasised by the fact that within-chromosome estimates are negatively correlated with the estimates of between-chromosome interaction. Thus, the + and − effects observed here could equally be due to flaws in the estimation, and not necessarily to properties associated with the genetic system. The results can fit both explanations reasonably well, and until exact estimates of the additive and within-chromosome effects are possible, there is no means of knowing which explanation to prefer. Not until within-chromosome effects can be dissociated from effects of the background, in the way made possible by the use of reciprocal substitutions, may such difficulties be overcome.

(ii) Genome interaction

The three genomes of the hexaploid wheat—A, B and D—are equally represented in the diallel cross discussed here. Comparisons between them are, therefore, possible.

TABLE 7

The number of times chromosomes from each of the genomes participate in significant genetic constants for the characters yield per plant, ear number per plant, height per plant and 500 grain weight

Type of Genetic constant	Character	Genome		
		A	B	D
Between-chromosome interaction	Yield . .	4	6	8
	Ear Number .	6	7	9
	Height .	2	8	4
	Grain Weight	4	5	1
	Total . .	16	26	22
Within-chromosome interaction	Yield . .	—	2	—
	Ear Number	1	2	1
	Height .	—	3	—
	Grain Weight	—	1	—
	Total . .	1	8	1
Additive chromosome effects .	Yield . .	2	1	1
	Ear Number	1	1	1
	Height .	2	—	—
	Grain Weight	3	2	3
	Total . .	8	4	5

The number of chromosomes participating in significant estimates can be classified in terms of the three genomes and the frequencies so obtained are shown in table 7. With so small a sample of significant estimates it is obviously not possible to say conclusively that one or other of the genomes participates to a greater or lesser extent than the others. Certain trends are apparent, however, and three of these that seem to summarise the main features of the table will be discussed.

First, the incidence of within-chromosome interaction is most frequent in the B genome. Even admitting that pleiotropy may reduce the total number of chromosomes involved for the four characters, the predominance of the B genome is still striking. Secondly, the A genome has the lowest overall frequency of chromosomes participating in between-chromosome interaction. This is consistent for all the characters studied except grain weight, where the A and B genomes give high values compared with the D genome. By contrast, the A genome has the highest incidence of additive effects. Thirdly, the B genome has the highest overall number of between-chromosome interactions, although only two of the characters, height and grain weight, exhibit this preponderance. Both yield and ear number have most interactions involving the D genome.

TABLE 8

The classification of between-chromosome interactions with respect to genomes and homoeologous groups

Character	Within Genomes			Between Genomes					
				Non-homoeologous			Homoeologous		
	AA	BB	DD	AB	AD	BD	AB	AD	BD
Yield	—	1	1	1	3	2	—	—	1
Ear Number	—	1	1	2	3	2	—	1	1
Height	—	2	—	1	1	2	—	—	1
Grain Weight	—	1	—	2	—	—	1	1	—
Total	—	5	2	6	7	6	1	2	3

Other pointers may be obtained by classifying the between-chromosome interactions even further. Thus, chromosomes in the same or different genomes may participate in between-chromosome interaction. Furthermore, between-genome interactions may occur between chromosomes belonging to the same or different homoeologous groups. Three classes of interaction may consequently be described between chromosomes, namely, within genomes, between genomes— between-homoeologous groups, and between genomes—within homoeologous groups. The frequencies of interaction for these three classes and for the four characters studied are shown in table 8. Altogether 32 between-chromosome interactions were observed of which seven occurred within genomes, 19 between genomes and between homoeologous groups and six between genomes but within a homoeologous group. Now, the maximum number of first order interactions possible

F

where nine chromosomes are involved is 36, of which nine involve within genomes, 18 between genomes non-homoeologously, and nine between genomes-homoeologously. The expectation of between-chromosome interaction for the three major classes delimited in the table on a purely random basis is consequently 1:2:1. There is therefore a reasonable fit when all the characters are combined between this random expectation and the observations, although there is some evidence suggesting differences between characters. Thus, grain weight has a greater number of homoeologous interactions than the other three characters.

TABLE 9

The number of times chromosomes from each of the homoeologous groups studied participated in significant genetic constants for the characters: yield per plant, ear number per plant, height per plant and 500 grain weight

Type of genetic constant	Character	Homoeologous Group		
		1	3	7
Between-chromosome interaction	Yield	3	6	9
	Ear Number	4	7	11
	Height	2	5	7
	Grain Weight	4	3	3
	Total	13	21	30
Within-chromosome interaction	Yield	—	1	1
	Ear Number	—	2	2
	Height	1	1	1
	Grain Weight	1	—	—
	Total	2	4	4
Additive chromosome effects	Yield	2	2	—
	Ear Number	1	2	—
	Height	1	1	—
	Grain Weight	3	2	3
	Total	7	7	3

An important feature which also emerges from this type of classification is the high frequency of within-genomic, between-chromosome interactions which concern the B genome chromosomes. Indeed, the higher value obtained for the B genome is similar to the high frequency of within-chromosome interactions which occur in this genome. The significance of this will be referred to later.

(iii) Group Interactions

Just as the localisation of significant genetic effects to the genomes may be of interest, so also will a localisation to particular homoeologous groups. Substitutions for the chromosomes of three homoeologous groups, 1, 3 and 7, were used in the diallel cross and the number of chromosomes participating in additive and interactive effects and occurring in each group is depicted in table 9. A number of features is immediately apparent. Apart from the character grain weight, which appears to have a totally different genetic architecture by this classification also, an excess of interactions is found in group 7. Indeed, unlike the genomic classification, for the three characters—yield, ear number and height—similar distributions occur for both the between- and within-chromosome interactions. There is, therefore, some evidence that for the chromosomes sampled, those of one homoeologous group play a predominant role so far as gene interactions are concerned in the control of three of the characters studied.

The experiment involving a diallel cross between substituted lines has shown that chromosome interactions can easily be recognised by this method. The location of these interactions with respect to genomes and homoeologous groups shows certain trends, so that further refinements to the genetic architecture may possibly be discerned. The next section is intended to provide further evidence to support the results which have been described.

5. THE HOPE SUBSTITUTED SERIES GROWN IN A NUMBER OF ENVIRONMENTS

When all the substituted lines and the recipient and donor varieties are grown together, then, as shown earlier in section 2, estimates of the summed additive effects (A) and summed between-chromosome interactions (I) for all the chromosomes can be obtained. An experiment, using the Hope substitution lines, was consequently undertaken to investigate the variations in these two components of the genotype in three different environments. Three controlled environment chambers were used in which temperatures of $15°C.$, $20°C.$ and $25°C.$ respectively were maintained.

Measurements were made on a number of characters and the data from five of them are considered here. Considerable variation occurred both between different substituted lines within a particular environment and also between environments. However, of great interest is the magnitude and sign of the two estimates $\hat{A}$ and $\hat{I}$. These have been obtained for the five characters and are given in table 10.

A number of points can be made from a consideration of the table. First, nearly all the A estimates are negative, although for the character spike length negative effects may not be the most prevalent. Conversely, the I estimates are largely positive. Indeed, the results from the summed estimates obtained by this method agree closely with the summation of the individual chromosome estimates obtained in the diallel cross. Secondly, there would appear to be no consistent effect of temperature on either the A or the I estimates for the five characters studied. Nor would there appear to be any consistently positive correlation of either type of estimate with increasing temperature. This contrasts with some of the results obtained in *Arabidopsis* and *Drosophila*, where an increase in gene interaction has been observed at higher temperatures for a number of characters (Langridge, 1962). The maximum temperature used here, however, was lower than the maximum temperature used with these two organisms, so that a positive correlation may not emerge until higher temperatures are tried. Thirdly, and perhaps of greater interest, is the environmental sensitivity of the two estimates. The estimates of *A*, although variable, would appear to be less variable than the estimates of *I*. This may follow from the higher standard errors involved in the estimation of *I* compared with *A*. Alternatively, it may result both from the higher sensitivity of interactions than of additive genetic effects to environmental differences, and from the

TABLE 10

Summed additive and summed between-chromosome interaction estimates for five characters at three temperatures

Character	A estimates			I estimates		
	15° C.	20° C.	25° C.	15° C.	20° C.	25° C.
Ear Emergence . . .	−7·81	−5·94	−7·06	+12·22	−3·66	+1·10
Tiller number . . .	−0·19	−1·63	−1·06	−1·68	+1·81	+3·10
Fertile tiller number .	+0·25	−0·25	−0·07	+5·27	+2·94	+2·03
Height	−6·50	−6·20	−4·30	+1·10	+26·40	+9·00
Spike length . . .	−11·31	+4·22	+4·13	−1·11	+23·45	+13·34
Total	−25·56	−9·80	−8·36	+15·80	+50·94	+28·57

predominance of between- over within-chromosome interactions in the expression of many characters. Indeed, intuitively, a large influence of environment on interactions may be expected, if only because mutually adjusted systems, such as occur in interactions and have evolved under particular environmental conditions, are more likely to break down under adverse environmental changes than are the more direct additive effects. Other observations support this conclusion. For example, in extensive studies with maize the partitioning of the genotypic variance into the two components—general and specific combining abilities—taken over locations and years has shown that the latter component is much more variable (Matzinger *et al.*, 1959). Likewise, the predominance of between- over within-chromosome interaction is supported from the observations made in the diallel cross in which the average magnitude of the interactions occurring between chromosomes is higher than the average of the estimates of within-chromosomes interaction. The higher sensitivity of genes involved in interaction would consequently seem a not unreasonable conclusion to make from the observations presented here.

It follows, therefore, that genes which are concerned with interaction may be detected by their larger contributions to environmental variation. Likewise, chromosomes which are concerned with interaction may be detected by their effect on environmental variation. This statement is likely to be true on average, and only on average, since complications must arise as a result of those genes—either additive or interactive—which are relatively infrequent and do not behave in an average manner. Thus, to take a single chromosome or a small sample of chromosomes and show that such samples are highly variable may be misleading, since there is no telling whether such a limited sample

TABLE 11

Environmental variances for the three genomes and for the five characters studied

Character	Genome		
	A	B	D
Height	2·66	6·93	3·41
Fertile Tillers . . .	0·03	0·22	0·29
Tillers	0·46	1·08	0·65
Spike Length . . .	5·19	8·67	5·50
Ear Emergence . . .	0·73	1·38	5·39
Total	9·07	18·28	15·24

is biased in favour of those genes which are abnormal so far as environmental sensitivity is concerned. However, where larger samples of the genotype are concerned, such as say a whole genome of wheat, then such distortions are less likely to be so large and the environmental variances observed are more likely to result from the variation of gene interaction with environment. Some localisation of chromosome interaction may consequently be obtained from intervarietal chromosome substitutions by studying the variation produced by groups of chromosomes over a range of environments.

This approach has been carried out with the Hope substitutions already mentioned. Genomic environmental variation has been calculated for each of the three genomes, A, B, and D, and the variances so obtained are shown in table 11. To increase the efficiency of this method, it is possible to calculate the variances based on the estimates of additivity for each chromosome rather than the actual observed values. Thus, it will be recalled that the comparison of each substitution line with the recipient variety gives rise to a value which combines the additive component of the chromosome being substituted and the interactive components of this chromosome with the remaining twenty wheat chromosomes. This provides a far better measure of individual chromosome interaction and environmental variation than if the observed values are used as the basis for calculating the variance. The comparison of each substituted line with the recipient has, therefore, been used in calculating the variances shown here.

A comparison of the environmental variances obtained from the three genomes of the Hope substitutions with the participation of these three genomes in chromosome interaction, described in the previous section, demonstrates that a remarkable agreement occurs between the two. For the two characters which are comparable in the two sets of data—height and the number of fertile tillers—a very good positive

TABLE 12

Environmental variances for the three homoeologous
groups and for the five characters studied

Character	Homoeologous Groups		
	1	3	7
Height	2·21	4·09	4·17
Fertile Tillers . . .	0·08	0·05	0·04
Tillers	0·56	0·30	0·42
Spike Length . . .	3·47	4·47	1·12
Ear Emergence . . .	1·27	0·73	3·77
Total	7·59	9·64	9·52

correlation exists between the number of chromosome interactions per genome and the environmental variance. For height it is the B genome which has the highest environmental variance and the highest frequency of chromosome interactions, whereas for the number of fertile tillers it is the D genome which is the highest in both experiments. Further, if the complications introduced by pleiotropy and the inclusion of other characters are ignored, there is again a remarkable agreement between the total number of chromosome interactions for the four characters presented in the diallel cross and the total of all the environmental variances for the five characters studied in this experiment. In both cases, the B genome has the highest value, followed by the D genome and with the A genome falling far behind them both. The apparent weaknesses in both experiments—the small sample of chromosomes used in the diallel crosses, and the possible confounding of additive with interactive effects or vice versa in the second—are lessened to some extent by the very close agreement between them.

This is not to say that such weaknesses should be ignored. Indeed, the environmental variances obtained for the three homoeologous groups, 1, 3 and 7, indicate that where the number of chromosomes considered is only three, that is the three chromosomes from one homoeologous group, the agreement between the two experiments is not as good (table 12). Thus, although the variances obtained for height and the total show some agreement, the variances calculated for the number of fertile tillers is completely the reverse of the number of chromosome interactions occurring in each of the groups. The agreement between the two experiments is, therefore, not so good when homoeologous groups are considered. This may follow, as was suggested earlier, from the distorting effects which may arise when small samples of chromosomes are used for detecting genetic interaction. Alternatively, it may point to a purely chance explanation of the close agreements which have been shown to occur between experiments. At the moment, however, it is more useful and more reasonable to assume that the close agreements which occur for the three genomes are meaningful and provide good evidence about the distribution and behaviour of genetic factors within the Hope substituted series.

6. THE GENETIC ARCHITECTURE OF WHEAT

The present results were obtained from two experiments using one substituted series—in which pairs of Hope chromosomes separately replace pairs from Chinese Spring. The genetic effects described and their allocation to chromosomes, genomes and homoeologous groups, in a way which illuminates the genetic architecture of wheat, may consequently be specific to the particular genotypes employed and provide no grounds for generalisation. Nevertheless, the results obtained from the use of the Hope substitutions may be usefully discussed and related, where possible, to some of the established facts of polyploid evolution. Indeed, by adopting this approach, it may be possible to establish a body of evidence which will demonstrate a coherent picture of the genetic architecture of wheat, even though the evidence from the Hope substitutions on their own can provide little scope for generalisation.

(i) Dominance and the initial polyploids

Some ideas about the genetic architecture of a polyploid such as wheat can be obtained by considering a simplified model of a tetraploid in which there are only two groups of homoeologous chromosomes. Further, let it be assumed that there is conventional dominance between alleles at corresponding loci of the homoeologous chromosomes. In this situation with only dominance occurring in the initial diploid, interaction of a duplicate nature will take place between homoeologous chromosomes in the tetraploid. Suppose, however, that gene interaction of a complementary nature occurs between genes of the two non-homologous chromosomes of the initial diploids. The tetraploid will then produce a similar interaction, but this will be influenced by the dominance relations between the alleles involved. Thus, if the genes in one diploid are dominant to the alleles found in the other, then gene interaction will be confined to one genome in the initial tetraploid and the remaining alleles from the other diploid parent will effectively become hypomorphs. However, if the dominance relations are equally distributed between the diploids, then interaction will occur between the genomes and between homoeologous groups.

Within-chromosome interactions may be visualised in a similar way. Thus, if two genes are found on each chromosome and interactions occur between them, then their behaviour in the tetraploid will depend on the dominance relationships of the alleles within a homoeologous group. If both dominant alleles are found on one chromosome, then non-allelic interaction will remain confined to that chromosome in the tetraploid. If, however, dominance is equally distributed then the within-chromosome interaction of the diploid will no longer remain, and a between-chromosome interaction will result which will occur between genomes and within a homoeologous group.

It is apparent from this discussion that the incidence and magnitude of the dominance effects which occur between the alleles contributed by the donor diploids are of paramount importance. With dominance alone and no non-allelic interaction in the diploids, between-chromosome interaction in the polyploid will be entirely between genomes and will occur homoeologously. This group of interactions will be further increased if within-chromosome interactions of a non-allelic nature occur in the diploids. A substantial number of interactions should consequently occur within the homoeologous groups in a polyploid such as wheat. Indeed dosage and interactive effects between duplicate and triplicate loci have often been suggested as an explanation of the failure of formal genetic analysis in wheat.

The present observations suggest, however, that of the three principal types of between-chromosome interaction, those between homoeologous chromosomes are least frequent. This may mean that dominance played a lesser role in the diploid ancestors than non-allelic interaction. Thus, dominance and therefore homoeologous interaction will be infrequent in the polyploids. Alternatively, and perhaps more likely, it may imply that the initial homoeologous interactions in the raw polyploid have been modified and lost as a result of mutation and selection. In this context, it is perhaps of significance that the observations found in the diallel experiment indicate that for a character such as grain weight, which has probably not been subjected to extensive selection, the ratio of homoeologous to other forms of interaction is much higher (table 8). Further, it may be expected that the D genome, since it is the most recent genome to be added in the evolution of the hexaploid wheat, will show a slightly greater number of homoeologous interactions than the A and B genomes. This follows, since selection has presumably had less time to operate after the formation of the hexaploid than is the case with the tetraploid. Indeed, if grain weight is excluded from the list of characters studied, then only homoeologous interactions involving the D genome are found.

(ii) The predominant effect of the B genome

It will be recalled that in both experiments, chromosome interactions predominantly involved B genome chromosomes. In the diallel experiment, the B genome contained the greatest number of significant within-chromosome interactions and the highest frequency of interactions between chromosomes belonged to this genome. Similarly, the B genome was found to have the highest environmental variance in the second experiment. Now, of the three probable diploid ancestors of hexaploid wheat, *Triticum monococcum*, *Aegilops speltoides* and *Aegilops squarrosa*, the donors of the A, B and D genomes respectively, it is the B donor alone which is likely to have a different genetic architecture. This follows from the fact that *Ae. speltoides* is the only outbreeder among the three (Zohary and Imber, 1963). An important genetic feature of outbreeding and one which is completely absent from inbreeding systems is the significance of dominance. Among outbreeders, as argued by Fisher (1930), natural selection will alter the phenotypic expression of heterozygotes towards the phenotype of the most favourable homozygote. Dominance of one allele over another will consequently occur. In inbreeders, however, these conditions are not fulfilled. Homozygous combinations of genes are the most frequent, so that although fortuitous dominance may occur, the circumstances by which the dominance of one allele over the other can evolve are never met. This is no great disadvantage to the inbreeder under favourable conditions; however, in a polyploid, formed by the combination of a diploid inbreeder and a diploid outbreeder, different conditions arise. Once again, homozygous genotypes will be formed as a result of chromosome doubling in the hybrid, but dominance will now be important, since similar alleles occur on different chromosomes within a homoeologous group. Under such circumstances it is not difficult to imagine that the alleles which are contributed by the outbreeder are more likely to be dominant than those which derive from the inbreeder.

It has been shown earlier in this section that, where dominance for the genes responsible for non-allelic interaction is confined to one of the diploids, interactions will occur either within chromosomes or between chromosomes from the genome contributed by this diploid. In other words, the non-allelic interactions of the initial diploid will tend to remain. In the case of wheat, an excess of within- and between-chromosome interactions involving chromosomes of the B genome was observed and, following the reasoning presented above, might have been expected. There is therefore reasoned support for the overriding effect of the B genome on chromosome interaction observed in the

Hope substitutions. The generality of such an effect on the hexaploid wheats is consequently more than just a possibility.

(iii) The significance of the D genome

One further feature in the origin of hexaploid wheat may also be of value in discussing its genetic architecture. This refers to the addition of the D genome in the formation of the hexaploid species from the AB tetraploid. Now, it has been suggested that, in the formation of the tetraploid, the outbred nature of the B genome contributor would have led to a bias in favour of alleles from this genome being recognised in the phenotype, whereas alleles in the A genome would have less effect. The D genome, since it derived from a natural inbreeder, may be supposed to behave similarly to the A, and therefore, the B genome may once again predominate. Indeed, gene interaction summed over all the characters follows this pattern—the B genome is predominant. However, for characters which, it may be argued, were under directional selection at the time the polyploid was formed, this situation is not necessarily true. Thus, it may be envisaged that the formation of the hexaploid depended on a satisfactory combination of a tetraploid species of *Triticum* with a diploid donor. A similar process may also be envisaged in the formation of the tetraploid from two diploid species. In both cases, gene interactions which modified the expression of a character towards the direction of selection would be preferred. In the tetraploid, however, selection would be concerned with those interactions which occurred between genomes A and B only; whereas at the hexaploid level selection would be involved with those interactions which occurred between the two already associated A and B genomes and the more recently added D genome. Consequently, the frequency of between-chromosome interactions which involve the D genome are likely to be more frequent than those involving either the A or the B genomes. But this will follow only for those characters under directional selection. For other characters, this situation would not arise and the predominance of the B genome would still emerge. The predominance of the B genome, therefore, reflects past selection pressures involved in the evolution of dominance at the diploid state rather than those present at the time of polyploid formation; the high frequency of D genome interactions demonstrates, on the other hand, the influence of selective forces operating only when the polyploid was formed.

The two characters, yield and ear number, in the diallel experiment, and the number of fertile tillers and ear emergence in the second experiment, show an excess of genetic interaction which concerns the D genome. All these characters are likely to be involved with directional selection at the time of polyploid formation and, according to the reasoning presented here, should give a predominance of D genome interactions. On the other hand, for the other characters studied, directional selection may well have been absent at the formation of the polyploid, so that the predominance of the D genome in chromosome interaction need not be expected in these cases.

To some extent, therefore, the reasoning which has been elaborated here agrees favourably with the observations which have been made using a single substituted series, that of the variety Hope into Chinese Spring. No doubt, the model which has been described will turn out to be a rather naïve representation of the genetic architecture of wheat. Indeed, the possibility that the origins of the polyploid wheats were polyphyletic has been ignored in the account which has been given. A monophyletic origin has been assumed, and although this seems probable (Riley, 1965), it must be emphasised that quite different interpretations of the evidence could be made if this assumption should prove false. The conclusions which have been made must therefore be considered as providing some of the concepts from which more definite ideas may develop. Further substituted series are required, not only into Chinese Spring but into other genetic backgrounds, so that the sampling of the wheat genotype may have a much firmer foundation. The genetic architecture of wheat, which has been observed perhaps only sketchily in the experiments described, may then be more fully portrayed.

7. SUMMARY

1. The development of appropriate aneuploid material in a polyploid species allows the substitution of single pairs of chromosomes between varieties. This enables chromosomes to be assayed with respect to their control of quantitative characters. At present, the requisite aneuploids are available in wheat, *Triticum aestivum*.

2. Some of the methods which can be applied to the inter-varietal substitutions of wheat are described.

3. The results derived from the F_1 generation of a diallel cross involving nine substitutions of chromosomes from the variety Hope into the variety Chinese Spring are given. Four characters, yield, ear number, 500 grain weight and height, were studied. Estimates of the genetic parameters for (1) additive chromosome effects, (2) within-chromosome interactions and (3) between-chromosome interactions are presented for each of these characters.

4. For yield, ear number and height, the estimates of the significant between-chromosome interactions are nearly all positive. By contrast, within-chromosome interactions are almost all negative. It is suggested that, if the complications of correlated estimation are ignored, this observation may be explained by supposing a homozygous genetic balance within chromosomes and a heterozygous balance between chromosomes. This could follow from the inbred, polyploid nature of wheat. On outcrossing, heterozygous combinations of genes result, so that the homozygous genetic balance within homologues is broken down. This is not so, however, for between-chromosome relationships, because a heterozygous balance exists.

5. The notion of a genetic balance for a character pre-supposes that directional selection has occurred. For characters not subjected to directional selection, genetic balance need not occur. It is argued that 500 grain weight, the only character not to show the features of genetic balance, is a character which has been subjected to stabilising rather than directional selection.

6. The distribution of chromosome interactions for the four characters with respect to chromosomes of the A, B and D genomes of the hexaploid wheat is described. The characters overall show an excess of chromosome interactions concerning the B genome. For the characters yield and ear number, however, an excess of D genome chromosomes participate in between-chromosome interaction.

7. An experiment using the 21 possible substitutions of chromosomes from the variety Hope, grown in three temperature environments, demonstrates a positive correlation between genome-environmental variation and the genome distribution of chromosome interactions described in the diallel experiment. It is argued that the genome-environmental variation is particularly influenced by chromosomes and genes involved in genetic interaction.

8. It is suggested that the predominant effect of the B genome on chromosome interactions may stem from the outbred nature of the probable B donor, *Aegilops speltoides*. The donors of the A and D genomes are both inbreeders, so that it may be argued that alleles contributed by the B donor would be dominant to those from the two inbreeders at the time of polyploid formation. The initial expression of this dominance may still be observed, in part, in the material studied.

9. The excess of D genome chromosomes involved in between-chromosome interactions for the characters yield and ear number may be explained in terms of directional selection for these two characters at the time of hexaploid formation. Selection would then be concerned with those interactions which occurred between a single genome, D, and two already associated genomes, A and B. Between-chromosome interactions are, therefore, likely to involve D genome chromosomes more often than those from the A and B genomes.

Acknowledgments.—It is a pleasure to acknowledge the helpful criticism and encouragement given by Dr. G. D. H. Bell and Dr. Ralph Riley. Valuable assistance was also given by Mr. A. J. Worland, Mr. T. E. Miller and Mr. H. A. Torrens.

8. REFERENCES

ALLARD, R. W., AND JAIN, S. K. 1962. Population studies in predominantly self-pollinated species. II. Analysis of quantitative genetic changes in a bulk-hybrid population of barley. *Evolution, 16*, 90-101.

BREESE, E. L., AND MATHER, K. 1960. The organisation of polygenic activity within a chromosome in *Drosophila*. II. Viability. *Heredity, 14*, 375-399.

DARLINGTON, C. D. 1939. *The evolution of genetic systems.* Cambridge Univ. Press, London.

FISHER, R. A. 1930. *The genetical theory of natural selection.* Oxford.

HAYMAN, B. I. 1958. The separation of epistatic from additive and dominance variation in generation means. *Heredity*, *12*, 371-390.

HAYMAN, B. I., AND MATHER, K. 1955. The description of genic interactions in continuous variation. *Biometrics*, *11*, 69-82.

JAIN, S. K., AND ALLARD, R. W. 1960. Population studies in predominantly self-pollinated species. I. Evidence of heterozygote advantage in a closed population of barley. *Proc. Nat. Acad. Sci., Washington*, *46*, 1373-1377.

JOHNSON, L. P. V., AND AKSEL, R. 1959. Inheritance of yielding capacity in a fifteen-parent diallel cross of Barley. *Canad. J. Genet. Cytol.*, *1*, 208-265.

KUSPIRA, J., AND UNRAU, J. 1957. Genetic analysis of certain characters in common wheat using whole chromosome substitution lines. *Can. J. Plant Sci.*, *37*, 300-326.

LANGRIDGE, J. 1962. A genetic and molecular basis for heterosis in *Arabidopsis* and *Drosophila*. *Amer. Naturalist*, *96*, 5-27.

LEWIS, K. R., AND JOHN, B. 1963. *Chromosome Marker*. Churchill, London.

MATZINGER, D. F., SPRAGUE, G. F., AND CLARK COCKERHAM, C. 1959. Diallel crosses of maize in experiments repeated over locations and years. *Agronomy Journal*, *51*, 346-350.

RILEY, R. 1965. Cytogenetics and the evolution of wheat. In *Essays on crop plant evolution* (ed. J. B. Hutchinson). Cambridge Univ. Press, London.

RILEY, R., AND LAW, C. N. 1965. Genetic variation in chromosome pairing. *Adv. Genetics*, *13*, 57-114.

SEARS, E. R. 1953. Nullisomic analysis in common wheat. *Am. Naturalist*, *87*, 245-252.

SEARS, E. R. 1954. The aneuploids of common wheat. *Res. Bull. Mo. agric. Exp. Stn.*, *572*.

UNRAU, J., PERSON, C., AND KUSPIRA, J. 1956. Chromosome substitution in hexaploid wheat. *Canad. J. Bot.*, *34*, 629-640.

ZOHARY, D., AND IMBER, D. 1963. Genetic dimorphism in fruit types in *Aegilops speltoides*. *Heredity*, *18*, 223-231.

ESTABLISHING A MONOSOMIC SERIES IN
AVENA SATIVA L.*

R. C. McGINNIS

Department of Plant Science, University of Manitoba, Winnipeg, Canada

1. INTRODUCTION

POLYPLOID species, because they are the product of parents having some genetic relationship, exhibit gene duplication of many characteristics. Such complex inheritance is difficult to study by conventional genetic methods. For several years monosomic series have been established in tobacco, *Nicotiana tabacum* L. (Clausen and Cameron, 1944) and wheat, *Triticum aestivum* L. (Sears, 1954); and their value in detailed genetic and cytogenetic analysis of these polyploid crops has been clearly demonstrated. It is reasonable to assume that similar success can be achieved in other polyploids such as common oats, *Avena sativa* L.

Common oats is an allohexaploid species with 42 chromosomes, comprising three genomes, A, C and D (Rajhathy and Morrison, 1959). Although it is an important crop, ranking fourth in total world production, comparatively little cytogenetic work has been done on it, and only recently have attempts been made to produce a complete monosomic series. The purpose of this paper is to indicate the progress achieved in establishing such a series and to outline the cytogenetic information already obtained by aneuploid analysis.

2. SPORADIC CHROMOSOME LOSS

Because the loss of a chromosome or chromosome pair frequently affects the phenotype, the first investigations of aneuploids in oats were confined to obvious off-types within a population where an attempt was being made to explain abnormal behaviour. As early as 1927, Huskins reported the association of the fatuoid complex with a chromosome deficiency. In addition to their fatuoid phenotype, the nullisomics were dwarf, sterile and highly asynaptic. The chromosome responsible was heterobrachial and designated " C ". This work was subsequently confirmed by other investigators, particularly Nishiyama (1933), who also used a telocentric chromosome to associate the fatuoid suppressing gene (or gene complex) with the short arm of the chromosome and the gene for synapsis with the long arm. Philp (1935) obtained albino nullisomics in the F_3 of a cross, *A. sativa gigantica* × *A. fatua*, and concluded that a gene for chlorophyll formation was

---

* Contribution No. 89 of the Plant Science Department, University of Manitoba.

located on the missing chromosome which he designated the V chromosome. In the reciprocal of the above cross Philp (1938) obtained nullisomics with narrow leaves and demonstrated that a gene for leaf width is associated with the chromosome designated L. He further reported that the C, V and L chromosomes were different from one another.

Surprisingly, these few reports comprised the aneuploid literature until 1954 when Tegenkamp and Finkner again associated albinism with a nullisomic condition. McGinnis in 1956 (as reported by Welsh, 1960) examined normal and fatuoid plants in foundation lines of Garry, and found three monosomics and two trisomics but no association of the off-type with aneuploidy. Recently two other albino nullisomics have been reported (McGinnis and Taylor, 1961; McGinnis and Andrews, 1962) and the specific chromosomes identified.

3. METHODS OF PRODUCING MONOSOMICS

(i) *Spontaneous Occurrence*

Although monosomics had been found occurring spontaneously, as mentioned in the foregoing section, no attempt to collect the different ones was made until recently. Riley and Kimber (1961) screened 631 seedlings of Sun II by root-tip chromosome counts and obtained seven monosomics. They suggested that it should be possible to assemble a monosomic series by cytologically screening large populations for their natural occurrence. McGinnis (1962) determined chromosome numbers in 4,023 seedlings in foundation lines of Garry and found two nullisomics, 17 monosomics and five trisomics, representing a frequency of 0·6 per cent. aneuploids in the variety. Of these it was determined that all five trisomics were different from one another and that there were at least six, and possibly eight or more different monosomics. Hacker and Riley (1963) reported having obtained six nullisomics, 40 monosomics, four trisomics and three plants with telocentrics out of a total population of 3,453 seedlings of Sun II for a frequency of 1·5 per cent. aneuploidy. Hacker and Riley (1964, personal communication) consider that possibly 13 different monosomics have been obtained. The variety Condor is being investigated for aneuploids by Holden and Marenah (personal communication). They have isolated a number of aneuploids by conducting chromosome counts on morphologically abnormal seedlings from " light seed " fractions. Thomas (personal communication) has found a high frequency of monosomics in off-type plants in the variety Manod and is confident that there are several different ones in his collection.

A summary of the occurrence of spontaneous aneuploids in hexaploid oats is presented in table 1.

The results to date indicate that a monosomic series could be produced in any variety by screening sufficiently large populations for chromosome deficiencies. Such a series offers the obvious advantage of being homogeneous. The disadvantage is primarily that the frequency of aneuploids is low and considerable labour and time would very likely have to be expended to obtain them all. Furthermore the data of McGinnis (1962) indicate that certain chromosomes are lost more frequently than others, and this would add to the difficulty of assembling the series.

TABLE 1

Spontaneous aneuploids in hexaploid oats

Variety or species	Number and type of aneuploid*	Authority
Victory	1M, 1N	Huskins (1927)
A. sativa gigantica × A. fatua	1M, 1N	Philp (1935)
A. fatua × A. sativa gigantica	1M, 1N	Philp (1938)
Ukraine × Trispernia	1M, 1N	Tegenkamp & Finkner (1954)
A. byzantina	1N	Ramage & Suneson (1958)
Garry	3M, 2T	McGinnis (reported by Welsh, 1960)
R.L. 157 × Ripon	1M, 1N	McGinnis & Taylor (1961)
White Russian × Exeter	1M, 1N	McGinnis & Andrews (1962)
Garry	17M, 2N, 5T	McGinnis (1962)
Sun II	40M, 6N, 4T	Hacker & Riley (1963)
Condor	several M, ?N, ?T	Holden (personal communication)
Manod	at least 32M, ?N	Thomas (personal communication)

* M = monosomic, N = nullisomic, T = trisomic.

(ii) X-irradiation

The application of low doses of X-irradiation, from 150 r to 600 r, to pre-flowering panicles has proved to be a highly productive method of inducing aneuploids in oats. The action of the irradiation is to induce chromatid breaks in pre-gametic nuclei. With illegitimate chromatid reunion dicentric chromosomes are formed and lost in subsequent divisions, giving rise to deficient gametes.

Costa-Rodrigues (1954) irradiated the variety Missouri 04047 and obtained 20 monosomics from 279 seedlings (7·2 per cent.). In spite of the obvious success of this method, it was not until seven years later that further investigations were conducted, and these by Chang and Sadanaga (personal communication) who produced 53 monosomics by irradiating Cherokee. McGinnis (1962, unpublished) induced 8·9 per cent. monosomics in Garry by irradiating at 300 r and also obtained a number of plants with 39, 40 and 43 chromosomes. Subsequent investigation by Andrews and McGinnis (1964) using four levels of irradiation, 75 r, 150 r, 300 r and 600 r, on Garry and Rodney gave evidence that, although the aneuploid frequency increases with the dose, so also does the frequency of undesirable chromosomal rearrangements. For example, at metaphase I many 41-chromosome plants from 600 r had $19^{II}C^{III}$ and frequently other aberrations, whereas most of those induced from the 300 r treatment had $20^{II}1^{I}$. At the two lower levels, no obvious aberrations were present in the monosomics. It was concluded that the most efficient X-irradiation dose is 150 r or slightly above, where the yield of monosomics is quite high (4·3 per cent.) and the degree of chromosome damage is minimised. By this method a total of 100 monosomics or potential monosomics have so far been produced in Garry, and 60 in Rodney.

Rajhathy and Dyck (1964) investigated a number of methods of producing monosomics and concluded that X-irradiation was the most efficient. The variety Garry was treated at 300 r and 500 r and a high yield of aneuploids obtained (15·5 per cent.).

Judging by the present data, it appears most probable that all the different monosomics could be produced easily by this method, and in fact may be present among those already produced. The results of the irradiation work are summarised in table 2.

Although the method is highly productive, it would appear essential that these monosomics are backcrossed to the untreated variety a number of times to reconstitute the original genetic purity and to eliminate minor chromosomal alterations. On the basis of recent work conducted at our laboratory, it has become apparent that, even in those plants which have $20^{II}1^{I}$ at metaphase I, their progeny can have a karyotype altered from the normal.

(iii) Interspecific hybridisation

Although a great many interspecific hybrids have been studied in *Avena*, relatively little of the work was directed toward the production of monosomics. Rajhathy and Dyck (1964) have shown that monosomics are found in a high frequency in the progeny of pentaploid hybrids. However, they consider the method somewhat unsatisfactory, mainly because of the heterogeneity introduced through such wide crosses but also because of the probable presence of chromosome rearrangements which would cause sterility in aneuploid × normal hybrids. One further disadvantage of the method appears probable. Since all known polyploid species of *Avena* have the A genome in common (Rajhathy and Morrison, 1959), such interspecific hybrids could give rise to only 14 of the monosomics, those in the C and D genomes. The remaining seven would have to be produced by some other means.

(iv) Progeny of Haploids

In wheat, Sears (1939) obtained a number of different monosomics by pollinating a haploid with normal pollen and demonstrated this to be an excellent source of such aneuploids. In addition to yielding a high frequency of monosomics, the varietal purity is maintained. Haploids apparently occur rarely in oats. Rajhathy and Dyck (1964), in an attempt to obtain a haploid from twin seedlings, screened 672,000 seedlings in varieties of *A. sativa* and found 148 twins, of which 145 were

G

TABLE 2

Aneuploids induced in hexaploid oats by X-irradiation

Variety	Irradiation dose (r)	Number of progeny screened	Chromosome no. of aneuploids				Authority
			39	40	41	43	
Missouri 04047 .	300	279			20		Costa-Rodrigues (1954)
Cherokee .	300, 600				53		Chang & Sadanaga (personal communication)
Garry .	300	506	4	5	41	1	McGinnis (unpublished)
Garry .	75, 150, 300, 600	686	2	8	18	1	Andrews & McGinnis (1964)
Garry .	300, 500	200		11	34	2	Rajhathy & Dyck (1964)
Rodney .	75, 150, 300, 600	599	6	15	45	1	Andrews & McGinnis (1964)

2n—2n and three were 2n—3n. The first haploid in hexaploid oats has just been reported by Nishiyama and Tabata (1964) who examined twin seedlings in Kanota, a variety of *A. byzantina*, and obtained one haploid plant. The meiotic behaviour is of interest since the majority of pollen mother cells (84 per cent.) had no bivalents, and only two bivalents per cell was the maximum observed. The haploid was partially self-fertile and also outcrossed readily. It is highly probable that a number of monosomics will be found in these progenies which are at present under investigation by Nishiyama. Since *A. sativa* and *A. byzantina* cross readily, such deficiencies can easily be introduced into any desired variety.

4. IDENTIFICATION OF MONOSOMICS

A number of methods of identifying monosomics are available and have been applied with varying degrees of success. These include karyotype analysis, metaphase I analysis of 40-chromosome F_1 hybrids between different monosomics or nullisomics, variation in the pheno-typic expression of monosomics or nullisomics, variations in trans-mission rate of the univalent chromosome, and the fertility of mono-somics and nullisomics. On the basis of recent findings, however, the author has come to the conclusion that the only reliable method of identification is by cytological analysis.

The chromosomes of common oats exhibit considerable variation. Idiograms of the species have been prepared recently (Rajhathy and Morrison, 1959; Rajhathy, 1963), which show that it is possible to distinguish nine of the 21 chromosomes as follows: 1, 2, 3, 4, 8, 9, 10, 15 and 21 (fig. 1). The remaining 12 chromosomes can be divided into four groups according to their length and centromere position. These are 11 and 12; 7 and 16; 5, 6, 13 and 14; and 17, 18, 19 and 20. To identify these chromosomes, intercrosses to produce 40-chromosome hybrids are confined to representatives within groups. Those having $19^{II}2^{I}$ at metaphase I result from crossing between parents with different chromosome deficiencies, whereas those with 20^{II} are from parents deficient for the same chromosome. For the first two groups only one cross is required and for the last two groups six intercrosses in each are necessary to sort out the different monosomics.

Although certain differences in plant morphology can be attributed to different chromosome deficiencies, it would be unwise to use this criterion exclusively in sorting out the monosomics. In Garry, for example, five different plants of monosomic 21 were found spon-taneously in foundation lines (McGinnis, 1962). Among these plants there were three distinct monosomic phenotypes which would have been classified as different deficiencies had not critical karyotype analyses been conducted. Thus, if plant appearance alone were used for identification, one could collect a number of monosomics deficient for the same chromosome expecting them to be different.

It has been observed that different monosomics may exhibit a difference in the rate of the univalent transmission, and this has been used to some extent to distinguish them (Chang and Sadanaga, personal communication). Certain lines such as 15 and 21 can produce up to 64 per cent. nullisomics (McGinnis and Taylor, 1961; McGinnis and Andrews, 1962). Other lines produce a very low frequency or none. In most lines, disomics are obtained in a low frequency, an extreme case in Garry (a line of chromosome 15) producing only one disomic in 400 seedlings examined. Recent findings (Lin and McGinnis unpublished) have indicated that transmission rate may be of limited value in identifying monosomics. In two of the Garry lines monosomic for 21 and in three other varieties monosomic for this chromosome, a study was conducted to determine the transmission rate of selfed monosomics under a number of environmental conditions. One line of Garry transmitted the univalent at a very low rate, the other at a relatively high rate and both were dependent on the temperature at the time of meiosis in the parental monosomic plants. The other varieties also exhibited a variable transmission pattern indicating that

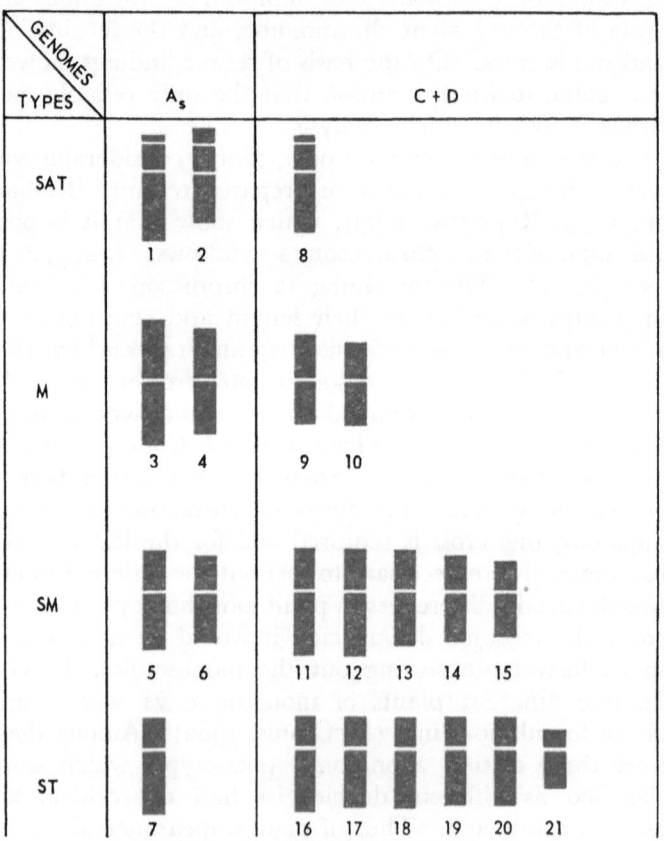

FIG. 1. Idiogram of *Avena sativa* L. (after Rajhathy, 1963. Reprinted by permission).

both environment and genotype are important in determining univalent transmission.

As a guide in distinguishing which chromosome is deficient, the degree of fertility of various monosomics and nullisomics may be used, but it is insufficiently precise for positive identification. Fertility is greatly influenced by the environment and monosomics are sensitive to minor variations in temperature and humidity. Thus under one set of conditions, a monosomic has been found to be highly fertile and under other conditions completely sterile.

5. PRESENT STATE OF MONOSOMIC INVESTIGATIONS

(i) Monosomic Series

As indicated in tables 1 and 2, a large number of monosomics have been obtained particularly in the varieties Garry, Rodney, Cherokee and Sun II. The major task confronting workers in this field at present is to sort out the different ones. In Garry, at least 10 different monosomics have been identified. These are monosomics 5, 6, 9, 10, 14, 15, 17, 18, 20 and 21. Large numbers of monosomic plants have not yet been studied and it is possible that most or all of the remaining 11 are among these. Identification of the different monosomics in Rodney is just under way. In Sun II, Hacker and Riley (personal communication) consider that possibly 13 different monosomics have been collected and Thomas (personal communication) is also confident that several different ones are present in the variety Manod. Chang and Sadanaga (personal communication) consider that they have seven different ones in Cherokee. There is little doubt that a series will be completed within a few years; indeed it would appear that series in several different varieties may be completed almost simultaneously.

(ii) Gene Locations

One of the major uses of monosomics is to associate genes with specific chromosomes. Numbers of such associations have been possible and they are presented in table 3.

Of particular interest is the demonstration that chromosomes 2, 15 and 21 each carry a gene for chlorophyll production. It is quite possible that these chromosomes constitute a homoeologous series, in which case chromosomes 15 and 21 would be in different genomes. Undoubtedly a great deal more information of this type will be forthcoming within the next few years.

6. DISCUSSION AND CONCLUSIONS

Because common oats and common wheat have the same chromosome number and each comprise three different genomes, it might be expected that they would exhibit similar responses to chromosome deficiencies. It is therefore somewhat surprising to observe the rather

TABLE 3

Gene-chromosome associations in A. sativa

Phenotype	Chromosome	Arm	Variety	Authority
albinism	2		Russell × 44-4	Dyck & Rajhathy (personal communication)
striated leaves	14	short	Garry	McGinnis & Sun (unpublished)
curled leaves	14	long	Garry	McGinnis & Sun (unpublished)
fatuoid	sub-median		Victory	Nishiyama (1933)
asynapsis	sub-median		Victory	Nishiyama (1933)
albinism	15		White Russian × Exeter	McGinnis & Andrews (1962)
albinism	15		Garry	McGinnis (unpublished)
side-panicle	15		Garry	McGinnis & Lin (unpublished)
thick-stems	15		Garry	McGinnis & Lin (unpublished)
kinky neck	20		Garry	McGinnis & Gauthier (unpublished)
albinism	21		R.L. 1574 × Ripon	McGinnis & Taylor (1961)
albinism	21	long	Rodney	Rajhathy (personal communication)
albinism	21		Rodney[5] × Exeter	McGinnis et al. (1963)

marked differences in the behaviour of their aneuploids. It would seem appropriate to emphasise the characteristics that appear to be peculiar to the oat aneuploids.

One of the more interesting observations is the very high frequency of nullisomics found in the progeny of certain of the monosomics (up to 64 per cent. as compared with 1 to 10 per cent. in wheat). Other lines produce a very low frequency or none. Presumably a low nulli-somic yield could indicate extreme certation or perhaps zygotic lethality of most nullisomics. Disomics are obtained in a very low frequency in most lines.

Wheat monosomics are usually fully fertile, whereas oat monosomics have a reduced fertility in all lines and almost complete sterility in certain lines of Garry. Some lines may prove difficult to maintain. Fertility appears to be greatly influenced by the interaction of environment and genotype. Many monosomics are extremely sensitive to minor variations in temperature and humidity perhaps indicating less gene duplication than would be expected in a hexaploid and lending support to the suggestion of McGinnis (1962) that hexaploid *Avena* may be much more diploidised than many polyploids. Perhaps it is for this reason that certain nullisomics are rarely or never produced. The diversity in chromosome morphology also indicates an evolutionary diploidisation, particularly when compared with wheat which has little variation between chromosomes. Even the meiotic behaviour of the hap-loid in Kanota (Nishiyama and Tabata, 1964), where 84 per cent. of cells had no bivalents and a maximum of only two bivalents was found in any cell, is different from that observed in wheat, where up to six bivalents (Krishnaswamy, 1939) have been reported, and further adds to the evidence that the species acts as a diploid in many respects.

The unique behaviour of the oat monosomics makes this species an interesting one to explore extensively. Within the last three years probably more cytogenetic work has been done on this crop than ever before. It would appear that this important food crop has finally caught the attention of numerous researchers and will now command its rightful position in the cytogenetic world. Undoubtedly the establishment of a monosomic series will aid greatly in solving many of the problems confronting workers at the present time.

7. SUMMARY

1. Only in the last five years have concerted attempts been made to produce a monosomic series in *A. sativa*. The chief hindrance has been the lack of a suitable source of these aneuploids.

2. Two methods of obtaining monosomics are proving promising: (1) screening large populations by root-tip chromosome counts for spontaneous occurrences; and (2) inducing monosomics by treating preflowering panicles with low doses of X-irradiation. Although monosomics are found in very low frequencies in untreated populations

they should be homogeneous. Irradiation gives a high yield of mono-somics but may also introduce undesirable chromosomal rearrange-ments and mutations which should be eliminated by backcrossing.

3. Nine monosomics can be identified by their distinct chromosome morphology. The remaining 12 chromosomes can be divided into four groups. Intercrosses are confined, therefore, to monosomics or nulli-somics within each group. Plant morphology, univalent transmission and fertility can aid in identification of different monosomics but are not always reliable.

4. At least 10 different monosomics have been obtained in Garry. A large number of monosomics have also been found in Sun II, Rodney, Missouri 04047, Cherokee, Manod, and Condor. It is probable that a complete series will be available in a number of varieties almost simultaneously.

5. Monosomics and nullisomics are proving valuable in associating genes with specific chromosomes.

6. The fact that oats exhibits a low frequency of pairing in a haploid, has an asymmetrical karyotype and responds to minor environmental variations suggests that this species is more diploidised than many polyploids.

Acknowledgments.—The author is grateful to the National Research Council of Canada for financial support.

8. REFERENCES

ANDREWS, G. Y., AND MCGINNIS, R. C. 1964. The artificial induction of aneuploids in *Avena*. *Canad. J. Genet. Cytol.*, 6, 349-356.

CLAUSEN, R. E., AND CAMERON, D. R. 1944. Inheritance in *Nicotiana tabacum*. XVIII. Monosomic analysis. *Genetics*, 29, 447-477.

COSTA-RODRIGUES, L. 1954. Chromosomal aberrations in oats, *Avena sativa* L. *Agronomia lusit.*, 16, 49-79.

HACKER, J. B., AND RILEY, R. 1963. Aneuploids in oat varietal populations. *Nature*, 197, 924-925.

HUSKINS, C. L. 1927. On the genetics and cytology of fatuoid or false wild oats. *J. Genet.*, 18, 315-363.

KRISHNASWAMY, N. 1939. Cytological studies in a haploid plant of *Triticum vulgare*. *Hereditas*, 25, 77-86.

McGINNIS, R. C. 1962. Aneuploids in common oats, *Avena sativa*. *Canad. J. Genet. Cytol.*, 4, 296-301.

McGINNIS, R. C., AND ANDREWS, G. Y. 1962. The identification of a second chromo-some involved in chlorophyll production in *Avena sativa*. *Canad. J. Genet. Cytol.*, 4, 1-5.

McGINNIS, R. C., AND TAYLOR, D. K. 1961. The association of a gene for chlorophyll production with a specific chromosome in *Avena sativa*. *Canad. J. Genet. Cytol.*, 3, 436-443.

McGINNIS, R. C., ANDREWS, G. Y., AND MCKENZIE, R. I. H. 1963. Determination of the chromosome arm carrying a gene for chlorophyll production in *Avena sativa*. *Canad. J. Genet. Cytol.*, 5, 57-59.

NISHIYAMA, I. 1933. The genetics and cytology of certain cereals. IV. Further studies on fatuoid oats. *Jap. J. Genet.*, 8, 107-124. 1933.

NISHIYAMA, I., AND TABATA, M. 1964. Cytogenetic studies in *Avena*. XII. Meiotic chromosome behaviour in a haploid cultivated oat. *Jap. J. Genet., 38,* 356-359.

PHILP, J. 1935. Aberrant albinism in polyploid oats. *J. Genet, 30,* 267-302.

PHILP, J. 1938. Aberrant leaf width in polyploid oats. *J. Genet., 36,* 405-429.

RAJHATHY, T. 1963. A standard karyotype for *Avena sativa*. *Canad. J. Genet. Cytol., 5,* 127-132.

RAJHATHY, T., AND DYCK, P. L. 1964. Methods for aneuploid production in common oats, *Avena sativa*. *Canad. J. Genet. Cytol., 6,* 215-220.

RAJHATHY, T., AND MORRISON, J. W. 1959. Chromosome morphology in the genus *Avena*. *Canad. J. Bot., 37,* 331-337.

RAMAGE, R. T., AND SUNESON, C. A. 1958. A nullisomic oat. *Agron. J., 50,* 52-53.

RILEY, R., AND KIMBER, G. 1961. Aneuploids and the cytogenetic structure of wheat varietal populations. *Heredity, 16,* 275-290.

SEARS, E. R. 1939. Cytogenetic studies with polyploid species of wheat. I. Chromosomal aberrations in the progeny of a haploid of *Triticum vulgare*. *Genetics, 24,* 509-523.

SEARS, E. R. 1954. The aneuploids of common wheat. *Res. Bull. Mo. agric. Exp. Stn., 472.*

TEGENKAMP, T. R., AND FINKNER, V. C. 1954. Inheritance of albinism in oats and the effects of carbohydrates on rust sori development of albino oat leaves. *Ann. Meeting Amer. Soc. Agron.*, pp. 74-75.

WELSH, J. N. 1960. Ten year report on oat breeding at the Cereal Breeding Laboratory, Canada Department of Agriculture Research Station, Winnipeg. *Cereal News, Special Edition.*

ATTRIBUTES OF INTRA- AND INTERSPECIFIC ANEUPLOIDY IN *GOSSYPIUM**

META S. BROWN

Texas A & M University, College Station, Texas, U.S.A.

1. INTRODUCTION

THE genus *Gossypium* is favourable material for the recovery and study of aneuploids, as it includes allotetraploids which readily tolerate duplications and deficiencies, and a series of diploids which can be combined with tetraploids to make allohexaploids. The latter provide a further source of aneuploids, especially trisomics involving diploid genomes with different degrees of relationship to *G. hirsutum*. Single and multiple monosomics, trisomics and tetrasomics have been derived from *G. hirsutum* and its hybrids with the diploid species (Beasley and Brown, 1943; Brown, 1949). The salient features of these types will be described as a means of elucidating chromosome differentiation in *Gossypium* species and in *G. hirsutum* in particular.

2. ORIGIN OF ANEUPLOIDS

(i) *Monosomics*

Monosomics of *G. hirsutum* have occurred spontaneously and following non-disjunction in cytologically aberrant stocks. They have been derived also from irradiated material. Details of the origin of 51 of the 62 monosomics now available have been published (Brown and Endrizzi, 1964). It was shown that monosomics occur spontaneously in normal *G. hirsutum,* such as commercial varieties, but occur more frequently in stocks of diverse genetic background, as has been shown in wheat (Riley and Kimber, 1961). In cytological preparations of meiotic metaphase in *Gossypium*, displaced bivalents, near one pole instead of on the equator, are observed frequently enough to account for the number of monosomics and trisomics of spontaneous origin which are found. The occurrence of monosomes due to chromosome loss following irradiation, or as a result of irregular chromosome distribution in structurally heterozygous stocks, is a common phenomenon. It is their recovery and transmission in species which tolerate monosomics that is significant.

(ii) *Intra-hirsutum trisomics*

Trisomics, like monosomics, sometimes appear spontaneously in *Gossypium.* One has occurred repeatedly in field populations of a

* Contribution No. 4850 from the Texas Agricultural Experimental Station, College Station, Texas. Part of the work done under Regional Cotton Project S-1.

commercial variety (Endrizzi, McMichael and Brown, 1963) and another has occurred at least twice in a genetic marker line (unpublished.) Trisomics also occur as a result of non-disjunction in structurally heterozygous stocks. The greater number, however, has been recovered following backcrossing of colchicine-induced 4(AD) autopolyploids of *G. hirsutum* and their 3(AD) derivatives. Usually several backcrosses to normal *G. hirsutum* are necessary to reduce multiple trisomics to $2n+1$. Early backcross plants having five or more extra chromosomes are sterile as a rule. This limits the crossing to plants having two to four extra chromosomes and thus simplifies the isolation of single trisomes.

(iii) Hybrid trisomics

Trisomics of interspecific origin have been recovered from colchicine-produced hexaploids combining *G. hirsutum* with diploid species. One or more species with any of the five types of genomes, A, B, C, D and E are represented. Following the initial backcross of the hexaploid, the segregating population derived from the pentaploid must be crossed repeatedly to the tetraploid parent to reduce the number of multiple trisomics to $2n+1$. As in aneuploids derived from autopolyploids of *G. hirsutum*, plants having more than four additional chromosomes are sterile.

Trisomics of both the A_1 and A_2 genome species, *G. herbaceum* and *G. arboreum*, are easily obtained. Reciprocal crosses have been made with A_2 pentaploids, and probably also succeed with A_1 pentaploids. From A_2 pentaploids, crossed as ovule parents, single and multiple trisomics, many of which are sterile, are recovered. Through the pollen, the transmission of extra chromosomes is limited to one or two, and the great majority of plants, though fertile, are $2n$. A large number of plants is required in either technique of crossing in order to obtain a random assortment of simple trisomics.

Pentaploids of *G. anomalum*, with the B_1 genome, can be crossed in either direction; and trisomics have been obtained both ways.

Trisomics involving chromosomes of *G. sturtii*, *G. raimondii* and *G. stocksii*, with the C_1, D_5 and E_1 genomes respectively, have all been derived from hybrid pentaploids used as female parents only. Of these three hybrids, only a limited number of trisomics was recovered with *G. sturtii* chromosomes, and their transmission is erratic.

Pentaploids of the D_2 species, *G. armourianum* and *G. harknessii*, are generally sterile as female parent. Hence only single and double trisomics were recovered when the pentaploids were used as pollen parent. Few trisomics of *armourianum* chromosomes were obtained; but many were recovered from *harknessii*, this pentaploid having been crossed more extensively.

Cytological, morphological and other aspects of these aneuploid forms will be described in separate sections.

(iv) *Intra-hirsutum tetrasomics*

Tetrasomics of *hirsutum* chromosomes and of chromosomes of the diploid species were obtained in two ways: by selfing trisomic plants and by intercrossing two trisomics which proved to involve homologous or homoeologous chromosomes. In many crosses a limited amount of pollen was used in order to give $n+1$ gametes an equal opportunity to function in competition with n gametes.

Tetrasomics of four *hirsutum* chromosomes were obtained by selfing, and of two other chromosomes of the tetraploid by intercrossing. In addition, three or more tetrasomics have resulted from natural selfing of trisomic plants under field conditions.

(v) *Hybrid tetrasomics*

Tetrasomics were recovered from six of the eight diploid genomes which yielded trisomics. Among the 21 tetrasomics recovered, three were the result of intercrosses of trisomics, one each among trisomics of *herbaceum*, *arboreum* and *harknessii*. The remainder, 18 in number, were the result of selfing trisomic plants. No tetrasomics were recovered from trisomics of *G. sturtii* and *G. armourianum* which were few in number.

(vi) *Double trisomics*

Although many multiple trisomics, including double trisomics, were recovered from auto- and allopolyploids of *Gossypium*, only those double trisomics originating from planned intercrosses of separate trisomic stocks will be considered. Their significance lies primarily in their contribution to chromosome identification and in their effect on chromosome pairing.

3. CYTOLOGICAL ASPECTS OF ANEUPLOIDS

(i) *Aneuploidy in relation to chromosome size*

Cytological verification of the chromosome number in all aneuploid plants was made by examination of aceto-carmine squashes of meiotic first metaphase of pollen mother cells. During these analyses attention has been directed to differences in metaphase chromosome size in *G. hirsutum*, in the diploid species and in species hybrids and their derivatives. Although absolute chromosome size is subject to variation owing to the degree of contraction, univalent chromosomes of *G. hirsutum*, as in polyhaploids of this species, can usually be designated large or small, with the exception of four or five which are medium in size. There is some gradation within each genome, but those of the B genome are large chromosomes approximately twice the size of the small D genome chromosomes. Large univalents in monosomic or trisomic plants are thus about the same size as small bivalents, although easily distinguished from them at metaphase by a difference in shape.

A tentative separation into A and D genomes can thus be made for monosomes and trisomes of *G. hirsutum* on the basis of size alone.

It has been pointed out that among the monosomics of *G. hirsutum*, those involving the large, or A genome, chromosomes are most frequently recovered (Kammacher, Brown and Newman, 1957; Brown and Endrizzi, 1964; Endrizzi and Brown, 1964). Of 62 monosomic chromosomes, 50 are large and 12 are small. The same phenomenon is true of trisomics. Of 35 trisomics from *G. hirsutum*, 23 involve large chromosomes and 12 involve small chromosomes. The significance of this difference in recovery will be considered more fully in the section on chromosome identification and in the discussion.

(ii) Chromosome pairing in monosomics and trisomics

Monosomes, having no homologues, can segregate only as univalents, except in infrequent instances when they misdivide and form telocentric fragments. In succeeding generations, the recovery of plants with unequal bivalents is evidence of such misdivision. Six or more plants with telocentric bivalents have been recovered from monosomics, both large and small. Among trisomic stocks also, unequal bivalents have been observed. These occur in $2n$ plants and in trisomic plants derived from pentaploids or other multiple trisomics, where there is ample opportunity for univalent misdivision.

Another aberrant feature of aneuploid cytology is the occurrence of non-disjunction involving chromosomes other than those involved in the original aberration. Four monosomics have been recovered from trisomic stocks, and two from other monosomics. Other aneuploids, not yet fully exploited, include $2n+1-1$ and $2n+2-1$ and similar types which have been recovered from pentaploids and early backcrosses.

Chromosomes present in triplicate vary in their pairing behaviour depending upon the degree of their relationship. Intra-*hirsutum* trisomics, entirely homologous, pair as trivalents in about 50 per cent. of the cells observed. These trivalents are usually V-shaped, perhaps as a consequence of the equal attraction and competition between homologous arms.

In hybrid trisomics, as in hexaploids and pentaploids, derived from species hybrids (Brown and Menzel, 1952), all degrees of pairing, ranging from that equal to intra-*hirsutum* trisomics to almost complete lack of pairing, are observed. Trisomics from the two A genomes of *herbaceum* and *arboreum* pair as trivalents with a frequency approximating that of pure *hirsutum*. Moreover, in several trisomics involving chromosomes having end arrangements different from *hirsutum*, the extra chromosome occurred as part of a multivalent, either a chain of five or a chain of seven. Chromosomes of the B_1 genome, despite a close relationship between the A and B genomes in diploid hybrids, have a low pairing frequency in triploids, pentaploids and trisomics involving *hirsutum* and *anomalum*. Trivalents are found infrequently. Many cells, however, have three univalents; which suggests either some degree of competition or a specific effect on chromosome pairing.

Homoeologous chromosomes of the A_h genome of *hirsutum* and the A_1 and A_2 genomes of *herbaceum* and *arboreum* are approximately equal in size and cannot be distinguished morphologically. Chromosomes of the B_1 genome are slightly larger than those of the A genomes, but certain distinctions between A and B homoeologues cannot be made on size alone. Chromosomes of the C and E genomes, however, are exceptionally large and can be easily distinguished from *hirsutum* chromosomes.

Trisomes resulting from the addition of chromosomes of the C_1 genome rarely pair with *hirsutum* chromosomes, and are conspicuous as long-armed univalents. In the limited number of trisomics which were studied, few cells had more than one univalent, that is unpaired *hirsutum* chromosomes were rare. Although C chromosomes pair with both A and D chromosomes in diploid hybrids, they offer little competition to pairing of *hirsutum* chromosomes.

In E_1 trisomics, trivalents are observed in few plants. In some plants, many cells have not one, but three to five univalents. In these cells the *stocksii* chromosome can be distinguished by its large size. The presence or absence of this slight asynapsis suggests a differential effect among the *stocksii* chromosomes.

In trisomics of *hirsutum* derived from D genome species, several degrees of pairing behaviour occur. In a few trisomics involving D_2 chromosomes, from *harknessii*, only univalents are observed, despite D_2D_h homoeology. In other trisomics, either trivalents or univalents occur, as would be expected. The relative frequency of univalents and trivalents varies among trisomics and may indicate individual differences between chromosomes.

In *hirsutum* trisomics with D_5 chromosomes, from *raimondii*, trivalents are commonly formed. As in some of the D_2 trisomics, trivalents involving D_5 often have an interstitial chiasma, forming a closed bivalent plus a side arm, rather than a V-shaped trivalent. The significance of these configurations with respect to chiasma frequency and localisation will be considered later.

One feature noticed in the progenies of trisomics is an increase in pairing of the trisome, as measured by the frequency of trivalents compared with univalents. Segregation for difference in pairing frequency is observed in progenies of trisomics, which indicates that some additional chromosomes of diploid origin remain intact and others have acquired segments of *hirsutum* chromosomes, which thereafter pair more readily. This increase in pairing has been observed particularly among trisomics involving B_1, E_1 and certain D_2 chromosomes, where pairing of the initial trisome was low. This interchange of segments between chromosomes of *hirsutum* and diploid species explains in part the phenotypic variation found among siblings.

(iii) *Chromosome pairing in tetrasomics*

Cytological analyses of two *hirsutum*-derived tetrasomics showed both 27^{II} and 25^{II} 1^{IV}. The other four tetrasomics each had a high proportion of cells with many univalents and attenuated one-chiasma bivalents. Comparison of all data, including date of collection, showed that asynapsis was most extreme during late July and August, when high temperatures and low soil moisture conditions prevailed. Cytological analyses of the same plants made before mid-July or during fall and winter under greenhouse conditions were more regular, with many cells with 27^{II} or 25^{II} 1^{IV}.

Tetrasomics of hybrid origin reflect the pairing observed in the parental trisomics. With one exception, which was slightly asynaptic, tetrasomic siblings from the two *herbaceum* tetrasomics had 27^{II} or 25^{II} 1^{IV} at meiosis.

Among six *arboreum*-derived tetrasomics one was highly asynaptic. Its tetrasomic progeny were more regular under greenhouse conditions, when 25^{II} 1^{IV} or 27^{II} were observed. In the remaining tetrasomics variable pairing was observed, including 26^{II} 2^{I}, 25^{II} 1^{IV}, 24^{II} 1^{VI}, and in some plants, 23^{II} 1^{VIII}. The significance of these multivalents will be clarified in the section on chromosome identification.

Tetrasomics derived from the addition of chromosomes of *anomalum* and *stocksii* to *hirsutum* paired primarily as 27^{II}, with occasional univalents and rare quadrivalents. Tetrasomic offspring of *anomalum*-derived tetrasomics differed in pairing behaviour; some were partially asynaptic, others formed quadrivalents frequently.

Initial tetrasomics derived from *harknessii*, four in number, varied in their pairing behaviour much as did the trisomic derivatives of this D species. In some, the chromosomes paired as 27^{II}, or 26^{II} 2^{I}; in others, quadrivalents were observed. As in progenies of trisomics, offspring of tetrasomics were more variable in pairing behaviour, with a higher frequency of quadrivalents and the occurrence of partial asynapsis in two lines, in particular in one $2n+4$ plant.

The four *hirsutum* tetrasomics derived from *raimondii* were more uniform in pairing behaviour than *harknessii* tetrasomics. Pairing as 27^{II}, 26^{II} 2^{I}, 25^{II} 1^{IV} was observed in all. Pairing in the offspring of the two fertile tetrasomics was similar. The one double tetrasomic, which was completely sterile, had pairing configurations of 28^{II}, 25^{II} 1^{IV} 2^{I}, and 27^{II} 2^{I}; no cells with 24^{II} 2^{IV} were observed.

In tetrasomic derivatives of *stocksii* the usual configuration was 27^{II}, with occasional univalents and one questionable quadrivalent.

As was observed in the progenies of trisomics, chromosome pairing in the offspring of many tetrasomics was more variable than in the parental aneuploids. This change in pairing behaviour is attributed to crossing-over, and segregation, between chromosomes of *hirsutum* and the diploid species.

4. MORPHOLOGICAL CHARACTERISTICS OF ANEUPLOIDS

(i) *Monosomics*

As described earlier (Brown and Endrizzi, 1964), loss of one chromosome from *hirsutum* usually causes a reduction in size of vegetative and fruiting structures. There are nevertheless certain differences between monosomic phenotypes which may be attributed to the two genomes. Among monosomics for A genome chromosomes of *hirsutum*, with the exception of chromosome A2, whose absence gives a small cleft boll, it is boll shape and proportion, rather than size, which is affected. The reduction in number of seeds and increase in number of motes or aborted ovules, cause boll distortion and a rough boll surface. Among the genome D monosomics thus far identified, the characteristic effect of chromosome loss is an overall reduction in boll size without distortion or change in shape.

(ii) *Trisomics and tetrasomics of* G. hirsutum

Trisomics and their corresponding tetrasomics, where obtained, affect the vegetative and fruiting organs in several ways. One phenotype, which has been observed repeatedly in different lines, is characterised by polyploid-like features: short stocky plant habit; rigid stem, short lateral branches; thick, dark green leathery leaves; broad, blunt buds; flower parts correspondingly short, broad or thick; and coarse deltoid bolls. Not all these morphological features are expressed in each of the trisomics or tetrasomics which show one or more of them, but they do tend to be associated. They are observed most often in trisomics involving large chromosomes, including K1, K15, K26 and three of more recent origin, including Acala A1 (Endrizzi, McMichael and Brown, 1963). One trisomic with a small chromosome, K3, and its sterile tetrasomic, showed these vegetative characters but had small bolls.

Another phenotype observed among *hirsutum* trisomics and tetrasomics is the opposite extreme. Plants are bushy, with many long, thin branches; small, narrow, thin, light-green leaves, 1-3 lobed; slender buds and flowers and small bolls. This phenotype is especially striking in trisomic K21—a large chromosome—and its tetrasomic. Some trisomics have vegetative and fruiting parts that contrast in size; for example K17, a medium-large chromosome, and K24, a very large chromosome, when added to *hirsutum* both give plants with small light-green, 3-lobed leaves, but large round bolls. Both trisomics yielded fertile tetrasomics with partial asynapsis at metaphase.

There is considerable variation in the morphology of the vegetative and fruiting parts of *G. hirsutum* depending upon age of plant, vigour of growth, light intensity and other factors. These variations sometimes obscure or confuse the phenotype due to specific trisomic chromosomes. It has been observed also that with selection some features are

⌄ost and others retained or modified. Backcrossing to isolate trisomics in a uniform genetic background, such as is now being practised in the monosomic study, would clarify which characters are specific to individual chromosomes.

(iii) Hybrid trisomics and tetrasomics

The many trisomics, recovered from hexaploids combining *G. hirsutum* with any of the eight diploid species named above, represent an array of aneuploids of great morphological variability. A detailed description of the genetic characteristics introduced into *hirsutum* from the diploid species is beyond the scope of this paper. Furthermore, when crossing-over and gene exchange occur freely, as in hybrids of *hirsutum* with the A and D genome species, the phenotype observed is not necessarily due solely to the extra chromosome. Hence, only brief reference will be made to known gene differences among the trisomics from the different species.

Genetic characteristics of diploid species which have found expression in trisomics include Petal Spot (brought in from *herbaceum*, *arboreum*, *anomalum*, *armourianum*, *harknessii* and *raimondii*), Yellow petal and Yellow pollen (from *herbaceum*, *arboreum*, *harknessii* and *stocksii*); Leaf shape (from *herbaceum*, *arboreum* and *harknessii*), Hirsute leaf and stem (from *anomalum*), Glabrous leaf (from *armourianum* and *harknessii*), Buff lint (from *harknessii* and *raimondii*) and Green seed coat (from *herbaceum*, *harknessii* and *raimondii*).

Not all genetic characteristics initially observed in trisomic plants were transmitted with the trisomic condition; for example, in a selfed progeny of one D_2 trisomic which initially carried Petal spot and Buff lint, $2n$ plants had Petal spot and tetrasomic plants had Buff lint.

Apart from the specific genetic characteristics brought in by chromosomes of diploids, the interaction of genes of the diploids in *hirsutum* background leads to extremes of morphological development. There are several pronounced morphological phenotypes which are characteristic of certain diploid genomes, or at least of those chromosomes which are most readily recovered in aneuploids. The polyploid-like phenotype observed in certain *hirsutum* trisomics and tetrasomics was found also in three trisomics with the extra chromosome derived from *arboreum*. Each of these gave fertile tetrasomic plants with rigid stems; large, dark green, folded leaves; blunt buds, and broad ruffled flowers. Leaves varied greatly in *arboreum*-derived trisomics, but another phenotype found among trisomics and emphasised in tetrasomics was characterised by narrow, one- to three-lobed leaves. Bolls also varied among trisomic lines, but all were large and sharply pointed as in *arboreum*. There was little or no reduction in boll size, as a large-bolled commercial variety of *arboreum* was involved in the initial hybrid.

In trisomics of *hirsutum-herbaceum* origin, modifications of many

H

parts of the plant were observed. One trisomic was distinguished by the *herbaceum* leaf shape in which the lobes are rounded, and the other by a small boll resembling that of *herbaceum*.

In *anomalum*-derived trisomics also, many features of the diploid parent were observed. In one tetrasomic the soft hirsute leaf, narrow 3-toothed bracteole, and ruffled petal were present; in the other, narrow bracts, narrow petals, Petal spot, and *anomalum*-like boll. In later generations some of these characteristics were lost.

From the hexaploid combining *hirsutum* and *sturtii*, the single and double trisomics studied were large vigorous plants without any phenotypic expression of the C genome. No tetrasomics were recovered.

Among trisomics of *hirsutum* derived from the two D_2 species, *armourianum* and *harknessii*, many characteristics of the diploid species were present, and variation was observed among siblings as well as between different trisomics. Slender stems and branches, small, rounded, 3-lobed leaves, glaucous stems and leaves, and small bolls were characteristic. Among the four tetrasomics from *harknessii* these features were accentuated, producing small plants with intense red, long thin glaucous branches, small but broadly-lobed leaves, small flowers, and small smooth-pointed bolls. Exceptional plants in progenies of two trisomics were polyploid-like with stiff stems and dense, thick, dark green leaves.

In trisomics and the four tetrasomics from the *hirsutum-raimondii* hexaploid, the most frequently observed effect of aneuploidy on plant habit was the development of a coarse, thick-stemmed plant, with short internodes and peduncles, and less deeply lobed leaves, in some lines folded or crumpled. At least four boll types were observed—long, deltoid, round and broad—but always large despite the small size of the boll in *raimondii*. The double tetrasomic, derived by selfing a double trisomic, was exceptionally polyploid-like, with short, thick stems, short lateral branches, very dark green leaves and short, blunt buds.

Trisomics originating from *stocksii* showed great variation in leaf shape and size, but leaf lobes were always pointed, never round as in the parental diploid. Leaves were folded or ruffled in many aneuploids, and bracteoles were broad, as in *stocksii*. The variation in boll shape was striking, there being four or more types in different trisomics: round, with and without cleft tip; broad and deltoid; elliptical or pecan-shaped; and extremely long and tapering, or lanceolate. Bolls, especially the lanceolate type, were extremely smooth and free of glands, in marked contrast to the heavily glanded, small round boll of *stocksii*. The deltoid boll was present in one tetrasomic line; the lanceolate boll, in two.

One feature observed in trisomics originating from both *raimondii* and *stocksii* was epinasty, a bending down of the leaf petioles and blades. This reaction was interpreted as an indication of incompatibility due to chromosome unbalance.

5. CHROMOSOME IDENTIFICATION
(i) *Trisomics and tetrasomics of* G. hirsutum

Many years ago a programme was initiated by D. T. Killough at the Texas Station to identify the chromosomes of *G. hirsutum* by the trisomic technique used so successfully in other genera. Many tests were made using nine of the qualitative genetic mutants then available. Although *Gossypium* species were among the first to be studied genetically, only two linkage groups were known before 1946. One stock carrying five dominant mutants, considered independent on the basis of early linkage tests, was used extensively.

After several years of testing involving hundreds of plants, a number of associations were found among the 25 trisomics and one or more of the mutant genes under study. One trisomic was linked with Okra leaf, and others with Petal spot, Brown lint and Naked seed; several trisomics showed linkage with two or more of these mutants. The many negative tests and several inconsistencies, especially the association of the same trisomic with two or more presumably independent characters, led to discontinuation of the study. It was concluded that the technique, which depends upon transmission of extra chromosomes through ovules only, was not applicable to *Gossypium*, where some transmission of extra chromosomes through the pollen occurs.

Subsequently, it was shown that three of these mutants, Petal spot, Brown lint and Naked seed, are all in linkage group I (Stephens, 1955). This linkage group has since been placed on chromosome A7, and Okra leaf, in linkage group II, by translocation studies (Brown, unpublished). It therefore seems justified to accept as valid the data identifying three trisomics as A7 and one as D15.

Meanwhile, a second approach to chromosome identification was made by cytological means. The series of translocation stocks now in use for monosomic study was not available during the period devoted to trisomic analysis. Hence the technique was limited to intercrosses between trisomics or tetrasomics, where available. Crosses were made between trisomic lines involving chromosomes equal in size, that is large × large, and small × small, in order to identify duplicate chromosomes and establish non-identity. Crosses were made also between large and small trisomics, to verify judgment on chromosome size, and demonstrate male transmission of extra chromosomes.

Because of limited transmission through the pollen, critical $2n+2$ plants were not recovered in all tests. Of 28 positive tests, only two combinations, both involving small to medium-sized, D-genome, chromosomes, gave tetrasomics, showing that the two trisomics of each cross were identical. Tetrasomic plants of both combinations were slightly asynaptic, as was the tetrasomic which was derived from one of the trisomics when selfed. An intercross involving one trisomic from each pair shown to be identical gave $2n+1+1$ plants which were also slightly asynaptic, but which occasionally had two trivalents,

H 2

showing non-identity. There are therefore at least two chromosomes of the D genome of *G. hirsutum* which cause asynapsis when present in excess.

Of the 27 combinations which produced double trisomics, 10 involved crosses between two large or A genome trisomics. One of these combinations produced partially asynaptic $2n+1+1$ plants. Both of the chromosomes involved produced partially asynaptic double trisomic plants in combination with one of the small chromosomes already known to cause asynapsis. As in the D genome, there are at least two chromosomes of the A genome of *G. hirsutum* which cause asynapsis in aneuploids.

Two different combinations, in which two large chromosomes were simultaneously trisomic, and in which one trisomic chromosome was common to both, had polyploid-like morphologies. Other double trisomics were more often less robust and less thrifty than disomic plants.

Twenty-one trisomic lines were involved in one or more of the 28 combinations which yielded positive data. The recovery of only two instances of duplicate trisomics among these indicated that the trisomic lines must include many different chromosomes of *G. hirsutum*. However, since the maximum number of positive tests with any one trisomic line was eight, it is probable that many duplicates, especially among the A genome chromosomes, may have escaped detection.

One trisomic-tetrasomic line of *G. hirsutum*, which has polyploid-like morphology, has been tested by crossing with translocation stocks and shown to have chromosome A1 in excess (Endrizzi, McMichael and Brown, 1963). Since this tetrasomic is sterile, it is probable that the earlier tetrasomics of *hirsutum*, which were fertile, did not concern chromosome A1. Conclusions based on data from monosomics, translocations, and hybrid trisomics of *arboreum* (see below) suggest that the *hirsutum* trisomics giving a polyploid-like phenotype may have chromosome A2 in excess.

(ii) Hybrid trisomics and tetrasomics

Only one test of a hybrid trisomic with a qualitative mutant gave positive results. A trisomic from *raimondii*, D₅, was linked with Red plant colour (R_1) on *hirsutum* chromosome D16.

Among cytological tests, the cross between the two *herbaceum* tetrasomics gave further tetrasomics, rather than double trisomics. Recovery of tetrasomics indicated that the A_1 chromosomes of the initial trisomic lines were identical, despite differences in plant morphology or phenotype. One cross between an A_1 and an A_2 trisomic gave double trisomics, indicating non-identity.

Among *arboreum* trisomics, one cross gave double trisomics, with two trivalents at MI. A second cross, involving two tetrasomics with polyploid-like morphology, gave further tetrasomics, proving identity of the two chromosomes. These two lines and their intercross all produced tetrasomic offspring in which pairing ranged from 27^{II}, 25^{II} 1^{IV}, 24^{II} 1^{VI} and 23^{II} 1^{VIII}. The recovery of multivalents of eight chromosomes indicates that the extra chromosome is one of the first three chromosomes of *arboreum* which form a ring of six in hybrids with *hirsutum* (Menzel and Brown, 1954).

A cross between the two *anomalum* tetrasomics, which differed greatly in morphology and phenotype, yielded a double trisomic. Similarly, the cross between the two *armourianum* trisomics produced a double trisomic. Among intercrosses of *harknessii* trisomics, the only positive test, which combined trisomics with many phenotypic differences, produced tetrasomics. An intercross between trisomics of *armourianum* and *harknessii* gave a double trisomic, indicating non-identity. Among intercrosses of *raimondii* trisomics, a cross between two with contrasting boll types gave a double trisomic, with two small univalents or trivalents of unequal size. Intercrosses of *raimondii* with *armourianum* and *harknessii* trisomics yielded double trisomics in two crosses, in which the D_5 chromosomes could be distinguished from the D_2 chromosomes by their smaller size. The double trisomic plant involving the very small fragment-like D_5 chromosome was dark red, like many D_2 trisomics. The smallest D_5 chromosome therefore is not the one carrying the anthocyanin suppressor.

No intercrosses were made between trisomics of *sturtii* or *stocksii*.

In intercrosses of hybrid aneuploids, more double trisomics than tetrasomics were recovered. This result indicates that many different chromosomes were represented among the trisomic lines tested.

(iii) *Monosomics*

Of 62 monosomic lines now available, 50 with large and 12 with small chromosomes deficient, 35 have been tested cytogenetically (Endrizzi and Brown, 1964, and unpublished). Tests with known chromosomes in translocations show that among 30 large deficient chromosomes, 27 involve four A genome chromosomes, as follows: one is A1, 9 are A2, 11 are A4 and 6 are A6. The remaining three large chromosomes involve *herbaceum* and *arboreum* chromosomes. The five identified monosomics for small chromosomes, whose D genome constitution was verified by crossing with *G. raimondii*, were D16 (two monosomics), D17 (two monosomics) and D18 (one monosomic).

Endrizzi (1963) tested the first six identified monosomics of *hirsutum* with three linkage groups, I, II and IV, marked by seven genes (Stephens, 1955) and with three additional mutants. With the exception of one chromosome, A6, which carried H_2 (Pilose, on linkage group IV), all the monosomics were independent of the ten mutants tested. Earlier, extensive tests of monosomic lines × mutant lines carried out by Dr. M. Y. Menzel and the author were all negative. Recently, unpublished data of the author and co-workers have identified monosomic D16 with linkage group III, R_1 cl_1.

6. DISCUSSION

The close relationship of the A and D diploid species to *G. hirsutum* has two drawbacks with respect to the analysis of specific individual chromosomes in trisomic stocks. Pairing between the introduced chromosome and those of *hirsutum* takes place readily, and so permits both crossing-over and a measure of segregation. In some trisomics this assortment may approach randomness. Hence, following one or more generations it is not certain that the original chromosome from the diploid species has not been replaced by its *hirsutum* homoeologue. Moreover, not only the trisomic chromosomes, but other chromosomes of the parental species are subject to some degree of crossing-over and segregation, beginning in the initial F_1 triploid and possible in each succeeding generation. Thus the structural and genetic integrity of neither the A nor D genome chromosomes can be maintained in trisomics of hybrid origin. Even when qualitative characters or other plant characteristics known to be properties of the diploid species are observed, it cannot be assumed that these are controlled by genes on the trisomic chromosome. This applies to chromosomes of both A and D species, to a greater or lesser degree, depending first upon the species and secondly upon the particular chromosome of the genome, since the relationship of the two A species and each of the D species to *G. hirsutum* is not the same.

The difference in the numbers of A and D chromosomes among the monosomes and trisomes warrants further consideration. Aneuploids of chromosomes of the A genome are recovered more readily than those of the D genome, and in many instances are more vigorous and long-lived. Although no specific viability factors have thus far been identified with individual chromosomes, the absence or the duplication of D chromosomes appears to have a more pronounced effect on plant viability. This is in line with the view that the A chromosomes have contributed the characters of agronomic importance to the tetraploid species. The D chromosomes therefore have been subjected to greater selection pressure for the maintenance of plant viability and fertility.

A factor which may be of great significance in the differential behaviour of A and D chromosomes is the difference in DNA composition. In a survey of the species of *Gossypium* recently extended to include the E genome, Ergle, Katterman and Richmond (1964) found a correlation between genome constitution and cytosine/5-methylcytosine ratios. The small chromosome species, with the D genomes, have larger C/5MC ratios than the large chromosome species. A possible correlation between chromosome size, amount of chromatic material, chiasma formation and C/5MC ratio has already been proposed (Brown, 1963). A similar correlation is suggested between DNA composition and the recovery of monosomes and trisomes of the A and D genomes.

The number of identified D chromosomes is as yet too few to permit conclusions to be drawn on differential recovery rates within the genome. In the A genome, however, certain chromosomes are recovered much more frequently as aneuploids than others. Two of the chromosomes recovered most frequently as monosomes, A2 and A4, are involved in the natural chromosome interchanges as well as numerous induced translocations. Trisomes for these chromosomes were not detected by genetic tests, as mutant genes are not known for either chromosome. Furthermore the frequency of recovery of monosomic and trisomic plants may not be parallel; thus three trisomics were identified with linkage group I, on chromosome A7, whereas corresponding monosomes have not yet been recovered even though translocations involving this chromosome are available for their detection.

Thus far, the cytological approach to chromosome identification has been more successful in *Gossypium* than genetic tests. Tests for association between genetic markers and cytological stocks have been predominantly negative. The tendency for mutant loci to be grouped in certain chromosomes, as shown by linkage studies, may be partly responsible for this situation. The imbalance between chromosomes with qualitative mutants and those recovered as monosomic and trisomic may be less marked as the additional mutants and aneuploids now coming into use are completely analysed.

7. SUMMARY

1. Monosomic plants of *G. hirsutum* have occurred spontaneously, and have been recovered from irradiated material and following segregation and non-disjunction in structurally heterozygous stocks.

2. Trisomic plants have been recovered from autopolyploids of *G. hirsutum*, and occur spontaneously in commercial or mutant stocks. They result also from non-disjunction in structurally heterozygous stocks.

3. Hybrid trisomics, or alien addition lines, have been obtained by backcrossing allohexaploids of *G. hirsutum* and each of the following diploid species, representing A, B, C, D and E genomes: *G. herbaceum*, *G. arboreum*, *G. anomalum*, *G. sturtii*, *G. armourianum*, *G. harknessii*, *G. raimondii* and *G. stocksii*.

4. Tetrasomics, both intra-*hirsutum* and hybrid, have been obtained by selfing and intercrossing trisomic stocks.

5. Intra-*hirsutum* trisomics, of the A and D genomes, give metaphase pairing of $25^{II} 1^{III}$ and $26^{II} 1^{I}$ in about equal frequency.

6. Hybrid trisomics reflect the degree of chromosome pairing observed in triploids and pentaploids of *G. hirsutum* and diploid species; thus metaphase pairing ranges from that observed in *hirsutum* trisomics, in trisomics involving A_1 A_2 D_2 and D_5 species, to little or no pairing of the extra chromosome in trisomics involving B_1, C_1 and E_1 species.

7. Most tetrasomics are fertile and partially stable; some are asynaptic when subjected to environmental stress.

8. Monosomic plants are distinguished by one or more of the following characteristics: smaller plant habit, smaller leaves, smaller or slightly distorted bolls.

9. Trisomic plants vary in morphology from those with diminution effects such as is seen in monosomics to the opposite extreme, with coarse, polyploid-like plant parts. In corresponding tetrasomics these features are accentuated.

10. In hybrid trisomics and tetrasomics, characteristics of the diploid species may be expressed in plant habit; leaf size, shape and texture; and flower and boll type. Transgressive expression of boll size and shape is especially marked.

11. In *G. hirsutum*, monosomics and trisomics involving the large A genome chromosomes are recovered more frequently than those for the small D genome chromosomes.

12. This and other differences observed in A- and D-genome aneuploids may be correlated with differences in DNA composition, D genomes having a higher cytosine/5-methylcytosine ratio than A genomes.

8. REFERENCES

BEASLEY, J. O., AND BROWN, M. S. 1943. The production of plants having an extra pair of chromosomes from species hybrids of cotton. *Rec. Gen. Soc. Amer.*, *12*, 43.

BROWN, M. S. 1949. Polyploids and aneuploids derived from species hybrids in *Gossypium*. *Proc. 8th Int. Con. Genet.*, Hereditas, Sup. Vol., 543-545.

BROWN, M. S. 1963. Chromosome differentiation in genomes of *Gossypium*. *Proc. 10th Int. Con. Genet.*, *1*, 132.

BROWN, M. S., AND ENDRIZZI, J. E. 1964. The origin, fertility and transmission of monosomics in *Gossypium*. *Amer. J. Bot.*, *51*, 108-115.

BROWN, M. S., AND MENZEL, M. Y. 1952. Polygenomic hybrids in *Gossypium*. I. Cytology of hexaploids, pentaploids and hexaploid combinations. *Genetics*, *37*, 242-263.

ENDRIZZI, J. E. 1963. Genetic analysis of six primary monosomes in *Gossypium hirsutum*. *Genetics*, *48*, 1625-1633.

ENDRIZZI, J. E., AND BROWN, M. S. 1964. Identification of monosomes for six chromosomes in *Gossypium hirsutum*. *Amer. J. Bot.*, *51*, 117-120.

ENDRIZZI, J. E., MCMICHAEL, S. C., AND BROWN, M. S. 1963. Chromosomal constitution of " Stag " plants of *Gossypium hirsutum*, Acala 4-42. *Crop Science*, *3*, 1-3.

ERGLE, D. R., KATTERMAN, F. R. H., AND RICHMOND, T. R. 1964. Aspects of nucleic acid composition in *Gossypium*. *Plant Physiology*, *39*, 145-150.

KAMMACHER, P. A., BROWN, M. S., AND NEWMAN, J. S. 1957. A quadruple monosomic in cotton. *J. Hered.*, *48*, 135-138.

MENZEL, M. Y., AND BROWN, M. S. 1954. The significance of multivalent formation in three-species *Gossypium* hybrids. *Genetics*, *39*, 546-557.

RILEY, R., AND KIMBER, G. 1961. Aneuploids and the cytogenetic structure of wheat varietal populations. *Heredity*, *16*, 275-290.

STEPHENS, S. G. 1955. Linkage in Upland cotton. *Genetics*, *40*, 903-917.

ASPECTS OF CHROMOSOME MANIPULATION:
A RÉSUMÉ*

D. U. GERSTEL and T. J. MANN
North Carolina Agricultural Experiment Station, Raleigh, N.C., U.S.A.

MANIPULATION of chromosomes is not new; it goes back to the early hybridisers of the eighteenth century,† or at least to Boveri's experiments with dispermic eggs of the sea-urchin (1907). Boveri's studies provided a convincing answer to a problem which had occupied him for many years; namely, whether the chromosomes of an organism are alike in their genetic endowment, as Weismann (1893) believed, or whether the chromosomes differ in the hereditary units they contain. If today we feel that we have the answer, we owe this, in no little measure, to the shrewd experimentation and imaginative conclusions of Boveri. His procedure utilised the unbalanced distribution of chromosomes occurring in the first cleavage division of eggs fertilised by two sperms. The resulting cells could be separated, and specific differences in their subsequent development observed; this led to the postulate of qualitative differences between chromosomes. (For a recent review of Boveri's contributions see Baltzer, 1962.)

Boveri thus produced the same conditions of unbalance for specific chromosomes as were obtained later in the trisomics of *Datura stramonium* (Blakeslee and Belling, 1924) and subsequently in a series of other plant species. In this symposium we have heard about trisomic tomatoes (Rick and Khush). Now it is also known that the unbalance generated by aneuploidy may be a matter of degree and its effects may not always be pronounced. Thus, in triploid hyacinths unbalanced pollen grains show no evidence of physiological differentiation, fertility is high, and aneuploid clones have become preferred horticultural varieties. Darlington and Mather (1944) concluded that each chromosome has a balance like that of the entire set. A similarity to Weismann's concept is striking, even though the latter was based on the wrong reasoning. The alternative view, that *Hyacinthus orientalis* ($2n = 16$) may be a polyploid with a basic chromosome number of 4 and thus buffered against the detriments of unbalance, appears improbable (Darlington and Mather, 1944; Feinbrun, 1938, and personal communication).

The genetic contents of chromosomes may also be studied in monosomic deficiency types. Bridges (1921) showed this many years ago in *Drosophila* for the minute fourth chromosome. Since then monosomics were obtained in other diploids, but only very rarely. Thus,

---

* Aided by grants from the National Science Foundation.
† Kölreuter's (1761-1766) backcrosses of hybrids between *Nicotiana rustica* (a tetraploid) and *N. paniculata* (a closely related diploid) are examples.

Rick and Khush (1961) produced haplo-11 of the tomato, but no off-spring could be recovered. Generally, only polyploids tolerate chromosome losses permitting experimentation with deficiency types. In several polyploid plant species monosomics have become valuable tools of analysis, much in the same fashion as trisomics; examples were presented in this symposium for cotton (Brown) and tobacco (Cameron); we also heard a report that a complete monosomic series may soon be available in oats (McGinnis). The nullisomics of wheat (Sears) are even more useful. Monosomics and trisomics can be employed for locating genes with major effects in particular chromosomes, only if contrasting alleles are available for analysis. In the absence of contrasting alleles, the gene contents of a chromosome cannot be read directly from the corresponding monosomic where there is no dosage effect; some genes may have the same effect in the hemizygote as in double dose. Thus, the flowers of haplo-G *Nicotiana tabacum* are not of lighter colour than those of the disomic, even though the G-chromosome contains a factor essential for the formation of pigment (Clausen and Cameron, 1944). Other genes in an allopolyploid may have become hypomorphs, and then it takes several of a kind to produce a fit phenotype. This situation was demonstrated by the compensating nullisomic-tetrasomic combinations discussed by Sears. The reasons may vary. In the first place, mutational diploidisation may have reduced gene efficiency. Secondly, the genes may have retained the potency they had in the diploid progenitors and become diluted in the polyploid. Thirdly, modifier systems may have arisen in the polyploid and reduced the total effect of duplicates or triplicates to the norm. It would be desirable to design more experiments which discriminate between these possibilities of polyploid evolution.

Some studies germane to this question are already available. We mention here only a few. In tobacco several investigators have studied the reduction of duplicate gene systems to a simple condition (Clausen and Cameron, 1950; Mann, Weybrew, Matzinger and Hall, in press). In tomatoes Sansome (1933) has demonstrated that genes which are fully dominant in diploids may have dosage effect in autotetraploids. Stephens (1951), Giles (1962) and Lee (in MS.) have compared several loci in tetraploid cottons with their homologues in the ancestral species. However, all these investigators were concerned with qualitative distinctions. In this symposium, Law has suggested a method of studying polygenic differences between varieties of common wheat. His plan is based on substitutions of both homologous chromosomes from one variety in appropriate nullisomics of a standard line. The method could also be applied to crops where only monosomics are obtainable, but the technical difficulties would be greater.

This symposium is also concerned with applications of chromosome manipulation to plant breeding. First, one might consider cases of " genome engineering ", in which the genomes are retained intact. A very interesting application has been presented to us by Peloquin and

Hougas for the utilisation of haploids of the potato.   The method is unique, being based on the essentially autopolyploid nature of *Solanum tuberosum*, the consequent fertility of polyhaploids, and on frequent parthenogenesis in this crop.   Were it not for the latter requirement, met with only rarely, the experiences of Peloquin and Hougas could perhaps be extended to the breeding of other autopolyploids (also rare), particularly alfalfa.   Other proposals to combine whole selected genomes have been made in bananas by breeding males as diploids and selecting the best combinations with commercial triploids which produce unreduced eggs (Simmonds, 1959).

Aneuploids too can be employed in breeding programmes.   I shall give only one example.   A novel use of trisomics for the maintenance of male-sterile lines for hybrid barley production has recently been suggested by Ramage and Tuleen (1964).   They propose using tertiary trisomics, which carry a pair of recessive male sterility alleles in normal chromosomes, and, in addition, an extra chromosome composed of segments from two non-homologues with a dominant fertility allele close to the translocation point.   All functional pollen and all diploid offspring produced by such a " balanced trisomic " will carry only the genes for male sterility.   Eggs may, however, transmit the extra chromosome with the fertility allele, and the line can be maintained from some of the offspring produced by selfing.

When plant breeders speak of chromosome manipulation they may think primarily of interspecific hybridisation.   Great efforts have been directed over the years towards derivation of alien substitution races. The usual purpose of such work is to introduce into a crop plant a certain alien chromosome segment carrying little more than a particular gene needed for improvement.   Thus, deleterious effects associated with foreign chromatin or chromosomal unbalance can be minimised. The topic was reviewed too recently (Stephens, 1961a; Mann, Gerstel and Apple, in press) for us to want to go into details again.   Instead, selected issues will be mentioned briefly.

The first concern of a breeder who wishes to employ foreign species is a taxonomic one of species relationships.   There are problems of crossability, chromosome number and fertility; but the extent of chromosome homology is almost equally important, for on it depends whether Mendelian-type backcross techniques can be employed, as they can when homology is more or less complete.   On the other end of the spectrum we find the situation where recombination does not occur between the chromosomes of two species at meiosis.   In hybrid derivatives of this kind chromosomes may be broken artificially by means of chemicals or by irradiation.   Since broken ends tend to fuse, and often in new combinations, the desired arrangement may be selected.   The feasibility of this approach was demonstrated first in a now classic work by Sears (1956).

The preceding shows that one needs to measure the extent of chromosome homology before launching an interspecific breeding programme. Several methods are available, but since they are all based on the *behaviour* of chromosomes it must be recognised that this behaviour may be subject to gene control. For example, chromosome associations at the meiotic metaphase in interspecific hybrids are quite often affected by genes with asynaptic or desynaptic effects, and the observer may underestimate the homology of the chromosomes of the species crossed. The inverse—overestimation of homology—will not happen, for non-homologous associations are resolved before metaphase. An interesting situation exists in the wheats where the chromosomes of different species have retained considerable homologies; yet the chromosomes are sufficiently differentiated for homologues to pair preferentially in allopolyploids, provided a mechanism is present which enhances differential affinity and minimises pairing between mere homoeologues. There is such a mechanism, and its genetic nature has been recognised in a series of papers from the laboratories of Sears and Riley, starting in 1958 (Sears and Okamoto, 1958; Riley and Chapman, 1958). By eliminating these genes, that is by employing appropriate nullisomics, one can obtain recombinations between chromosomes which normally do not pair. Riley, Chapman and Kimber (1959) have suggested using this procedure for interspecific gene transfers.

A variety of ways has been proposed by which to measure the extent of the cytological similarity or dissimilarity of chromosomes. Riley has shown us here a method designed to measure the affinities between homoeologues in polyploid wheat and between wheat chromosomes and those of a wild species. Gaul (1954) and Sficas and Gerstel (1962) have described statistical methods by which one may ascertain the maximal number of bivalents which a given hybrid is potentially capable of forming. A technique for discovering cryptic structural differentiation (Stebbins, 1950), based on the extent of recovery of alleles from the non-recurrent parent in backcrosses, has been developed by Stephens (1949). Rick and Khush have mentioned an application to *Lycopersicon*. Linkages between genes transferred *en bloc* from one species to another were found to be tighter than in the original parent; it is significant that these changes represented shifts of the cross-over region on the chromosome rather than reduction in total recombination, at least in *Gossypium* (Rhyne, 1958; Stephens, 1961*b*). These authors attributed the reduced cross-over frequencies in the transferred regions to structural differences; alternatively, localisation of crossing-over may also be under genetic control, for it has been shown that the localisation of chiasmata may be so affected. Thus, Rees (1955) has demonstrated dependence of chiasma localisation on the genotype in rye. In *Triturus cristatus*, a newt, chiasmata tend to be localised distally in the male (Callan and Spurway, 1951) but proximally in the female (Callan and Lloyd, 1960), although the reality of this difference still requires proof, since metaphase chromosomes were studied in the male but female oocytes in the lampbrush stage. Unfortunately, no germane studies have been done with hermaphroditic organisms where one could compare localisation in the male and female organs of the same individual. Such a study would explain whether chiasma localisation may also be under control of cytoplasm differentiated during development.

If indeed genic or extra-chromosomal factors, rather than chromosome structure, were determining the shifts in cross-over position in the *Gossypium* hybrids, one would expect some transferred chromosome blocks to be affected in one way and others in the opposite direction; by testing a large number of such blocks some decision between alternatives may be attainable.

Finally, the extent of differentiation between chromosomes of related species may be measured in their allopolyploid hybrids; cytologically, by counting the frequency of multivalent chromosome associations (Sarvella, 1958; Phillips, 1964); and genetically, by assessing the deviations from tetrasomic ratios among the offspring (Collins and Longley, 1935; Gerstel and Phillips, 1958; Shaver, 1962). Since the results depend on the extent of preferential pairing in the allopolyploid hybrids, a genetic mechanism like the one controlling homologous bivalent formation in wheat may interfere. Methods of detecting such controls were developed by the wheat geneticists mentioned. When some of these methods and other criteria were applied to *Nicotiana* and *Gossypium*, little evidence for genotypic interference with association between homoeologues was detected in allopolyploids of these genera (Gerstel, 1965). Therefore, their genetic output reflects directly the structural similarities or dissimilarities of the chromosomes which were tested.

Before concluding, I would like to mention briefly an area about which our ignorance is still great; this concerns the fate of genetic factors after they have been transferred from one species to another. One may point to the successes in interspecific breeding in which a high level of disease resistance was incorporated from wild species into cultigens, indicating that no particular problem existed in those cases. But only a few outstanding successes have been described in the literature, while the frequent failures have often remained part of the lore of breeders of specific crops. Because of an acquaintance with cotton and tobacco we are aware of some difficulties here. Incorporation of the fine lint characteristics of *Gossypium barbadense* into the better yielding *G. hirsutum* has been tried by many without success despite the similarity in chromosomal structure of the two (Stephens, 1961b). Efforts to incorporate the impressive lint strength of *G. thurberi* into the Upland crop (*G. hirsutum*) have proved very frustrating (Al-Jibouri, Miller and Robinson, 1958). In tobacco it has been possible to introduce various kinds of disease-resistance from wild species into burley, but transfers of the same genes into flue-cured types of *N. tabacum* were largely unsuccessful.

To a considerable extent, failure of a " clean " transfer of a gene from one species to another may be attributable to linkages between the " good " and " bad " genes and there is always hope that even tight linkages can be broken with patience and by appropriate techniques. One must realise, however, that the architecture of a species does not involve chromosome structure alone, but that intra- and interchromosomal balances, developmental channelling and relations between chromosomes and cytoplasm are also involved. Genes are not independent building-bricks which can be juggled around freely but dynamic triggers of developmental action; the timing and expression of action are stimulated by extraneous controlling systems. The components of such a system may be closely linked to the " structural gene " and the stimulus proceed along the chromosome, or the components may be situated on different chromosomes (McClintock, 1956). In the latter case the stimulus must travel through a suitable cytoplasm; the cytoplasm may also be implicated as a carrier of self-replicating units with a phenotypic effect (Wagner and Mitchell, 1964) or in alternate steady states of metabolic patterns (Novick and Weiner, 1957; Preer, 1957). It may well be that the intricate balance of such systems is placed in jeopardy in interspecific hybridisation, just as it may be put out of phase as a consequence of chromosome breakage within a species (McClintock, 1951). Indeed, examples exist where such processes may be suspected. Best known are the many cases where incorporation of the chromosomes from one species into the cytoplasm of another upsets sexual development and results in male sterility (Caspari, 1948; Duvick, 1959). Interspecific hybridisation often generates instability; in *Nicotiana*, for example, mutable genes, developmental disorders and chromosome aberrations have all been encountered in hybrid derivatives (Moav and Cameron, 1960). More recently developmental instability of floral pigmentation and immensely oversized single chromosomes in sporadic cells, as well as chromosome fragmentation, were found in hybrid derivatives from *Nicotiana tabacum* × *N. otophora* in our laboratory (Gerstel and Burns, 1963; 1965). Examples of instability following hybridisation in other genera could easily be added (Mangelsdorf, 1958; Walters, 1957). There is here neither the time nor the place to elaborate; it would also be premature to do so, but it appears probable that chromosome manipulation, inter- and intraspecific, can make important contributions in this area. The plant breeder too will need to watch developments.

REFERENCES

AL-JIBOURI, H. A., MILLER, P. A., AND ROBINSON, H. F. 1958. Genotypic and environmental variances and covariances in an Upland cotton cross of interspecific origin. *Agronomy J.*, *50*, 633-636.

BALTZER, F. 1962. *Theodor Boveri. Leben und Werk eines grossen Biologen*, 1862-1915. Wissenschaftliche Verlagsgesellschaft, Stuttgart.

BLAKESLEE, A. F., AND BELLING, J. 1924. Chromosomal mutations in the Jimson weed, *Datura stramonium. J. Hered.*, *15*, 195-206.

BOVERI, T. 1907. Zellen-Studien. Die Entwicklung dispermer Seeigeleier. Ein Beitrag zur Befruchtungslehre und zur Theorie des Kerns. *Jena. Z. Naturw.*, *43*, 1-292.

BRIDGES, C. B. 1921. Genetical and cytological proof of non-disjunction of the fourth chromosome of *Drosophila melanogaster*. *Proc. Nat. Acad. Sci., Wash.*, *7*, 186-192.

CALLAN, H. G., AND LLOYD, L. 1960. Lampbrush chromosomes of crested newts, *Triturus cristatus* (Laurenti). *Phil. Trans. Soc. Lond. Sci. B.*, *243*, 135-214.

CALLAN, H. G., AND SPURWAY, H. 1951. A study of meiosis in interracial hybrids of the newt, *Triturus cristatus*. *J. Genet.*, *50*, 235-249.

CASPARI, E. 1948. Cytoplasmic inheritance. *Adv. Genet. 2*, 1-66.

CLAUSEN, R. E., AND CAMERON, D. R. 1944. Inheritance in *Nicotiana tabacum*. XVIII. Monosomic analysis. *Genetics, 29*, 447-477.

CLAUSEN, R. E., AND CAMERON, D. R. 1950. Inheritance in *Nicotiana tabacum*. XXIII. Duplicate factors for chlorophyll production. *Genetics, 35*, 4-10.

COLLINS, G. N., AND LONGLEY, A. E. 1935. A tetraploid hybrid of maize and perennial teosinte. *J. agric. Res., 50*, 123-133.

DARLINGTON, C. D., AND MATHER, K. 1944. Chromosome balance and interaction in *Hyacinthus*. *J. Genet., 46*, 52-61.

DUVICK, D. N. 1959. The use of cytoplasmic male sterility in hybrid seed production. *Econ. Bot., 13*, 167-195.

FEINBRUN, N. 1938. A monographic study on the genus *Bellevalia* Lapeyr. II. Taxonomic-geographical part. *Palest. J. Bot., Jerusalem ser., 1*, 131-142.

GAUL, H. 1954. Asynapsis und ihre Bedeutung für die Genomanalyse. *Z. indukt. Abstamm.-u VererbLehre, 86*, 69-100.

GERSTEL, D. U. 1965. Evolutionary problems in some polyploid crop plants. *2nd Intern. Wheat Symp., Lund.* (In press.)

GERSTEL, D. U., AND BURNS, J. A. 1963. Variegation following hybridization between *Nicotiana tabacum* and *N. otophora*. *Proc. 11th Intern. Conf. Genet., 1*, 45.

GERSTEL, D. U., AND BURNS, J. A. 1965. Sporadic giant chromosomes in hybrids between two species of *Nicotiana*. *Proc. 1st Intern. Conf. on Chromosomes, Abstr.* (In press.)

GERSTEL, D. U., AND PHILLIPS, L. L. 1958. Segregation of synthetic amphiploids in *Gossypium* and *Nicotiana*. *Cold Spring Harbor Symp. Quant. Biol., 23*, 225-237.

GILES, J. A. 1962. The comparative genetics of *Gossypium hirsutum* and the synthetic amphiploid *Gossypium arboreum* L. × *Gossypium thurberi* Tod. *Genetics, 47*, 45-59.

KÖLREUTER, D. J. G. 1761-1766. *Vorlaufige Nachricht von einigen das Geschlecht der Pflanzen betreffenden Versuchen und Beobachtungen, nebst Fortsetzung 1, 2 und 3*. Ostwalds Klassiker der Exakten Wissenschaften, W. Engelmann, Leipzig.

LEE, J. A. The genomic allocation of the principal foliar-gland loci in *Gossypium hirsutum* L. and *Gossypium barbadense* L. (Manuscript.)

MANGELSDORF, P. C. 1958. The mutagenic effect of hybridizing maize and teosinte. *Cold Spring Harbor Symp. Quant. Biol., 23*, 409-421.

MANN, T. J., GERSTEL, D. U., AND APPLE, J. L. The role of interspecific hybridization in tobacco disease control. *3rd Inter. Scientif. Congr., Salisbury, S. Rhodesia.* (In press.)

MANN, T. J., WEYBREW, J. A., MATZINGER, D. F., AND HALL, J. L. Inheritance of the conversion of nicotine to nornicotine in varieties of *Nicotiana tabacum* L. and related amphiploids. *Crop. Sci.* (In press.)

McCLINTOCK, B. 1951. Chromosome organization and gene expression. *Cold Spring Harbor Symp. Quant. Biol., 16*, 13-47.

McCLINTOCK, B. 1956. Controlling elements and the gene. *Cold Spring Harbor Symp. Quant. Biol., 21*, 197-216.

MOAV, R., AND CAMERON, D. R. 1960. Genetic instability in *Nicotiana* hybrids. I. The expression of instability in *N. tabacum* × *N. plumbaginifolia*. *Amer. J. Bot., 47*, 87-93.

NOVICK, A., AND WEINER, M.   1957.   Enzyme induction as an all-or-none phenomenon. *Proc. Nat. Acad. Sci., Wash., 43,* 553-566.

PHILLIPS, L. L.   1964.   Segregation in new allopolyploids of *Gossypium.* V. Multivalent formation in New World × Asiatic and New World × Wild American hexaploids. *Amer. J. Bot., 51,* 324-329.

PREER, J. R., JR.   1957.   Nuclear and cytoplasmic differentiation in the Protozoa. In: *Developmental Cytology.* D. Rudnick, ed. Ronald Press Co., New York.

RAMAGE, R. T., AND TULEEN, N. A.   1964.   Balanced tertiary trisomics in barley serve as a pollen source homogeneous for a recessive lethal gene. *Crop Sci., 4,* 81-82.

REES, H.   1955.   Genotypic control of chromosome behaviour in rye. I. Inbred lines. *Heredity, 9,* 93-116.

RHYNE, C. L.   1958.   Linkage studies in *Gossypium.* I. Altered recombination in allotetraploid *G. hirsutum* L. following linkage group transference from related diploid species. *Genetics, 43,* 822-834.

RICK, C. M., AND KHUSH, G. S.   1961.   X-ray-induced deficiencies of chromosome 11 in the tomato. *Genetics, 46,* 1389-1393.

RILEY, R., AND CHAPMAN, V.   1958.   Genetic control of the cytologically diploid behaviour of hexaploid wheat. *Nature, 182,* 713-715.

RILEY, R., CHAPMAN, V., AND KIMBER, G.   1959.   Genetic control of chromosome pairing in intergeneric hybrids with wheat. *Nature, 183,* 1244-1246.

SANSOME, F. W.   1933.   Chromatid segration in *Solanum lycopersicum.* *J. Genet., 27,* 105-126.

SARVELLA, P.   1958.   Multivalent formation and genetic segregation in some allopolyploid *Gossypium* hybrids. *Genetics, 43,* 601-619.

SEARS, E. R.   1956.   The transfer of leaf-rust resistance from *Aegilops umbellulata* to wheat. *Brookhaven Symp. Biol., 9,* 1-22.

SEARS, E. R., AND OKAMOTO, M.   1958.   Intergenomic relationships in hexaploid wheat. *Proc. 10th Intern. Cong. Genet., 2,* 258-259.

SFICAS, A. G., AND GERSTEL, D. U.   1962.   Statistical analysis of chromosome pairing in interspecific hybrids. II. Applications to some *Nicotiana* hybrids. *Genetics, 47,* 1163-1185.

SHAVER, D. L.   1962.   Cytogenetic studies of allotetraploid hybrids of maize and perennial teosinte. *Amer. J. Bot., 49,* 348-354.

SIMMONDS, N. W.   1959.   *Bananas.* Longmans, London.

STEBBINS, G. L.   1950.   *Variation and Evolution in Plants.* Columbia University Press, New York.

STEPHENS, S. G.   1949.   The cytogenetics of speciation in *Gossypium.* I. Selective elimination of the donor parent genotype in interspecific backcrosses. *Genetics, 34,* 627-637.

STEPHENS, S. G.   1951.   Possible significance of duplication in evolution. *Adv. Genet., 4,* 247-265.

STEPHENS, S. G.   1961a.   Species differentiation in relation to crop improvement. *Crop Sci., 1,* 1-5.

STEPHENS, S. G.   1961b.   Recombination between supposedly homologous chromosomes of *Gossypium barbadense* L. and *G. hirsutum* L. *Genetics, 46,* 1483-1500.

WAGNER, R. P., AND MITCHELL, H. K.   1964.   *Genetics and Metabolism,* 2nd ed. John Wiley and Sons, New York.

WALTERS, M. S.   1957.   Studies of spontaneous chromosome breakage in interspecific hybrids of *Bromus.* *Univ. Calif. Publ. Bot., 28,* 335-447.

WEISMANN, A.   1893.   *The Germ-Plasm, a Theory of Heredity.* (Trans. W. N. Parker and H. Rönnfeldt.) Charles Scribner's Sons, New York.

AUTHOR AND REFERENCE INDEX

Authors of papers published in this volume are indicated by an asterisk. Page numbers in italics refer to the extent of an author's contribution; any references to other papers by that author within his contribution are not indicated separately.